FM 21-22
AFM 64-26
US ARMY FIELD MANUAL

SURVIVAL AT SEA

1950

CIVILIAN REFERENCE EDITION

UNABRIDGED HISTORIC MANUAL ON SURVIVAL AND RESCUE FROM BEING ADRIFT, SHIP EVACUATIONS, AND WATER- AND ISLAND-BASED ESCAPE

U.S. DEPARTMENT OF THE ARMY

Doublebit Press

www.doublebitpress.com | Cherry, IL, USA

Original content under the public domain; unrestricted for civilian distribution.
Originally published in 1950 by the U.S. Department of The Army.

A Part of the Military Outdoors Skills Series

Doublebit Press Civilian Reference Edition ISBNs
Hardcover: 978-1643891989
Paperback: 978-1643891996

The Military Outdoors Skills Series

Historic Field Manuals and Military Guides on Outdoors Skills and Travel

Military manuals contain essential knowledge about outdoors life, thriving while in the field, and self-sufficiency. Unfortunately, many great military books, field manuals, and technical guides over the years have become less available and harder to find. These have either been rescinded by the armed forces or are otherwise out of print due to their age. This does not mean that these manuals are worthless or "out of date" – in fact, the opposite is true! It is true that the US Military frequently updates its manuals as its protocols frequently change based on the current times and combat situations that our armed services face. However, the knowledge about the outdoors over the entire history of military publications is timeless!

By publishing the **Military Outdoors Skills Series**, it is our goal at Doublebit Press to do what we can to preserve and share valuable military works that hold timeless history about outdoors life, navigation, and survival. These books include official unrestricted texts such as army field manuals (the FM series), technical manuals (the TM series), and other military books from the Air Force, Navy, and texts from before 1900. Through remastered reprint editions of military handbooks and field manuals, outdoors enthusiasts, bushcrafters, hunters, scouts, campers, survivalists, nature lore experts, and military historians can preserve the time-tested knowledge and history about the skills and institutional procedures that were learned through hard lessons and training by the U.S. Military and our expert soldiers.

This book is an important contribution to self-sufficiency literature and has important historical and collector value toward preserving these traditions.

Although the editors at Doublebit Press are thrilled to have comfortable experiences in the woods and love our high-tech and light-weight equipment, we are also realizing that the basic skills taught by the old experts are more essential than ever as our culture becomes more and more hooked on digital technology. We don't want to risk forgetting the important steps, skills, or building blocks involved with thriving in the outdoors. This Civilian Reference Edition historic reprint is a part of a collection of military handbooks and field manuals that are essential contributions to the American self-sufficiency tradition despite originating with the military. In the most basic sense, these books are the collection of experiences by the great experts of outdoors skills and self-sufficiency life: our countless expert soldiers who learned to thrive in the backwoods, deserts, extreme cold environments, and jungles of the world.

With technology playing a major role in everyday life, sometimes we need to take a step back in time to find those basic building blocks used for gaining mastery – the things that we have luckily not completely lost and has been recorded in books over the last two centuries. These history hasn't been forgotten, it's just been shelved. *It's time to unshelve this history once again and reclaim the lost knowledge of self-sufficiency.*

Based on this commitment to preserving our self-sufficiency heritage, we have taken great pride in publishing this book as a complete, unabridged and original work. We hope it is worthy of both study and collection by outdoors folk in the modern era of self-sufficiency and traditional skills life.

Unlike many other photocopy reproductions of classic books that are common on the market, this historic Civilian Reference Edition does not simply place poor photography of old texts on our pages and use error-prone optical scanning or computer-generated text. We want our work to speak for itself, and reflect the quality demanded by our customers who spend their hard-earned money. With this in mind, each Historic Edition book that has been chosen for publication is carefully remastered from original print books, *with the Doublebit Civilian Reference Edition printed and laid out in the exact way that it was presented at its original publication*. We provide a beautiful, memorable experience that is as true to the original text as best as possible, but with the aid of modern technology to make as beautiful a reading experience as possible for books that are typically over a century old. Military historians and outdoors enthusiasts alike are sure to appreciate the care to preserve this work!

Because of its age and because it is presented in its original form, the book may contain misspellings, inking errors, and other print blemishes that were common for the age. However, these are exactly the things that we feel give the book its character, which we preserved in this Historic Edition. During digitization, we ensured that each illustration in the text was clean and sharp with the least amount of loss from being copied and digitized as possible. Full-page plate illustrations are presented as they were found, often including the extra blank page that was often behind a plate. For the covers, we use the original cover design to give the book its original feel. We are sure you'll appreciate the fine touches and attention to detail that your historic Civilian Reference Edition has to offer.

For outdoors, self-sufficiency, and military history enthusiasts who demand the best from their equipment, the Doublebit Press Civilian Reference Edition reprint of this military manual was made with you in mind. Both important and minor details have equally both been accounted for by our publishing staff, down to the cover, font, layout, and images. It is the goal of Doublebit Civilian Reference Edition series to preserve outdoors and self-sufficiency heritage, but also be cherished as collectible pieces, worthy of collection in any library and that can be passed to future generations.

FM 21–22—AFM 64–26

This manual supersedes FM 21–22, 25 April 1944

SURVIVAL AT SEA

United States Government Printing Office
Washington: 1950

DEPARTMENTS OF THE ARMY AND THE AIR FORCE

Washington 25, D. C., *1 JUNE 1950*

FM 21-22—AFM 64-26 is published for the information and guidance of all concerned.

[AG 300.7 (16 Jul 48)]

By order of the Secretaries of the Army and the Air Force:

Official:
EDWARD F. WITSELL
Major General, USA
The Adjutant General

J. LAWTON COLLINS
Chief of Staff
United States Army

Official:
L. L. JUDGE
Colonel, USAF
Air Adjutant General

HOYT S. VANDENBERG
Chief of Staff
United States Air Force

Distribution:

Army:

Tech Sv (2) except 55 (20); AFF (5); OS Maj Comd (5); MDW (5); A (20); CHQ (5); D (5) except Tng Div (200); R (5) Bn (3); C 5, 8 (1); 55 (2); FC (2); Sch (15) except 55 (200); USMA (2); PMS&T (1); SH (1); PE (100), OSD (2).

For explanation of distribution formula, see SR 310-90-1.

Air Force:

Zone of interior:

Hq USAF (10); Maj Air Comds (10) each; Sub Air Comds (5) each; Services (MATS) (5) each; Air Divisions (5) each; Wings (5) each; Air Force Bases (2) each; Groups (4) each; Squadrons (3) each; Air Force ROTC Units (2) each; Air Div, NGB (info purposes only) (5) each.

Overseas:

Maj Air Comds (10) each; Sub Air Comds (5) each; Wings (5) each; Air Force Bases (2) each; Groups (4) each; Squadrons (3) each.

CONTENTS

This manual supersedes FM 21–22, 25 April 1944

CHAPTER 1

INTRODUCTION

1. PURPOSE. The purpose of this manual is to provide basic and fundamental information on how to take care of yourself and what to do during emergencies at sea.

2. SCOPE. This manual covers shipboard safety measures and equipment, abandon-ship procedure, operation and navigation of lifesaving craft, measures for sustaining life while adrift, rescue procedure, and brief suggestions for instructors. Since a landfall may prove only temporary refuge, some hints on survival ashore are also given. Swimming and lifesaving techniques, treated in detail in FM 21–20, are described only briefly in this manual.

3. GENERAL PRINCIPLES. **a.** Survival at sea means the ability of a person to take care of himself and others during an emergency afloat. This manual contains the important facts a man should know about the sea and about survival methods. The greatest hazard that will be encountered while adrift at sea is anxiety over the unknown. Such anxiety is largely caused by lack of information on means of survival at sea which this manual presents. A knowledge of what to do during an emergency and how to do it may save many lives. The time to acquire that knowledge is now, before an emergency arises.

b. Common sense is also necessary, however, and a considerable degree of ingenuity may be required, for not all the equipment and facilities mentioned herein may be available. Even though the equipment is at hand, precautions cannot be ignored. If any article of equipment is not available in an emergency, make the best of the situation. The principles in this manual, and applied common sense and ingenuity, will go far toward insuring survival at sea. In addition, rescue crews will be looking for survivors.

CHAPTER 2

SAFETY ABOARD SHIP

Section I. SAFETY EQUIPMENT

4. GENERAL. a. Ship's safety equipment. The safety equipment aboard a troop transport, exclusive of fire-fighting apparatus, usually includes lifeboats, life rafts and floats, exposure suits, life preservers, line-throwing guns and breeches buoys, life nets, embarkation and emergency escape ladders, ropes, lights (life-preserver, portable emergency and flash types), whistles, jackknives, and life buoys. A number of these items, as well as additional equipment, are found in lifesaving craft. (See pars. 5–8.)

b. Individual safety equipment. A life preserver will be provided for every man boarding a transport. It may be issued to each individual or stored in an easily accessible place. A life-preserver light, whistle, pamphlet of safety rules, and a single-blade jackknife may also be issued.

5. LIFEBOATS. a. Description. The modern lifeboat is a double-ender or whale-boat type, the bow and stern being of the same or nearly the same shape. It may be made of wood or metal and may be self-righting and/or self-bailing, containing watertight compartments and hand-pumps. It may be motor-driven or propelled by oars, sails, a hand-operated propeller, or combinations thereof.

b. Nomenclature. Nomenclature of a hand-propelled metal lifeboat is indicated in figure 1.

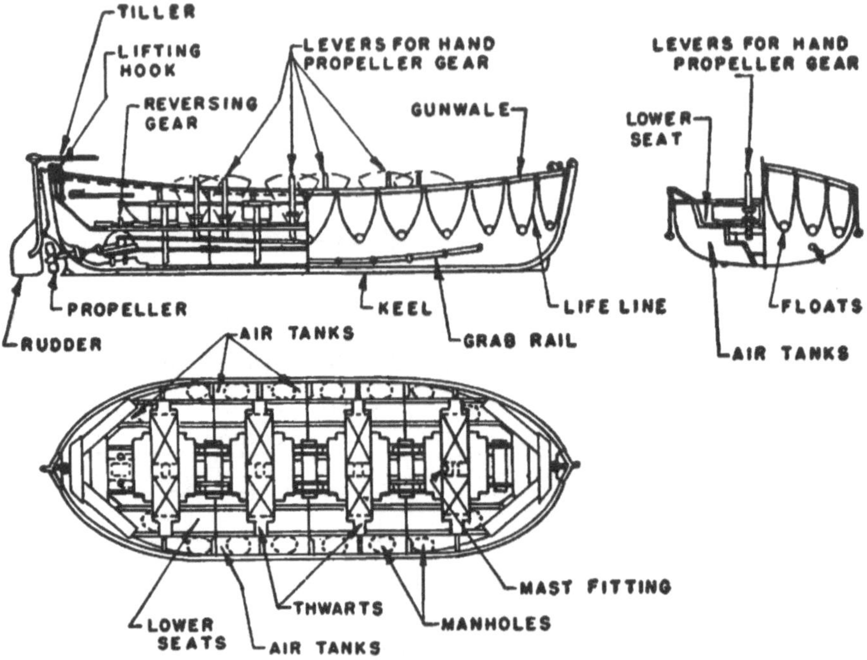

Figure 1. Nomenclature of hand-propelled metal lifeboat.

c. Equipment. The equipment listed below is required on lifeboats. Motor lifeboats are additionally equipped with motor and accessories, searchlight, radio installations, and two fire extinguishers, and they need not carry a mast, sails, or more than four rowing oars and one steering oar.

(1) Bailer.
(2) Two boathooks.
(3) One 2-gallon galvanized iron bucket.
(4) One 2-inch liquid compass.
(5) Twelve self-igniting red flares, 2-minute, in watertight case.
(6) One ditty bag with palm, needles, twine marlin, and marlinspike.
(7) Two enamel drinking cups.
(8) Two single-edge hatchets.

(9) One gallon illuminating oil.
(10) One lantern and oil for nine hours.
(11) One lifeline in 3-foot bights and a sieve float in each.
(12) Two life preservers.
(13) One locker for small gear.
(14) One mast, sails, and gear.
(15) Matches in waterproof case.
(16) Complement of oars and two spares.
(17) One 2¾-inch painter, length three times boat deck to water at light draft.
(18) Two plugs attached by chain to each drain hole.
(19) Steering oar and rowlock.
(20) Rudder and tiller or yoke.
(21) Sea anchor and oil cone.
(22) One gallon storm oil.
(23) Breakers, pump or spigot, containing one quart of water for each person.
(24) Provisions consisting of two pounds of bread and one can of condensed milk per person.

d. Wartime equipment. In time of war additional equipment may be provided as shown in the following list:

(1) Blankets.
(2) Charts.
(3) Curtain and hood.
(4) First aid kit.
(5) Fishing kit.
(6) Bullet hole plugs.
(7) Line.
(8) Signal flags.
(9) Signalling mirrors.

(10) Flashlight.
(11) Repair kit.
(12) Bible.
(13) Massage oil.
(14) Whistle, jackknife, and life preserver light for each person.

6. LIFE RAFTS. a. General. A life raft consists of air tanks or drums enclosed in a wooden frame. (fig. 2). The raft is reversible, with compartments for equipment that can be opened from either side. Life rafts are usually stowed so as to provide quick release directly into the water, and are arranged so they will have the best chance of floating free of the ship if there is no time to launch them. Life floats and rafts and designed primarily to provide flotation for survivors until rescue comes. Sails may be provided with some types of life rafts. When they are not provided, it is usually too exhausting even under normal conditions of wind and sea to attempt any headway in such craft. Strength should usually be conserved to keep afloat and not to make distance.

b. Equipment. Equipment required for life rafts is listed below:

(1) Lifeline (as in lifeboats).
(2) Painter (as in lifeboats).
(3) Four oars.
(4) Five rowlocks.
(5) Matches in waterproof case.
(6) Water (as in lifeboat).
(7) Bread (as in lifeboat).
(8) Twelve flares (as in lifeboat).
(9) Sea anchor and oil cone.

Figure 2. One type of life raft.

(10) One gallon storm oil.
(11) Two enamel cups.
(12) Self-igniting water light.

7. LIFE FLOAT. A life float is usually made of cork or balsa wood covered with canvas, with netting and slats for a deck (fig. 3). It may be launched directly overboard or can float free of a sinking ship. A white electric water light usually will be attached to the float by a lanyard, and at least two paddles will be lashed to the sides or the slats. Other equipment varies with the type of craft.

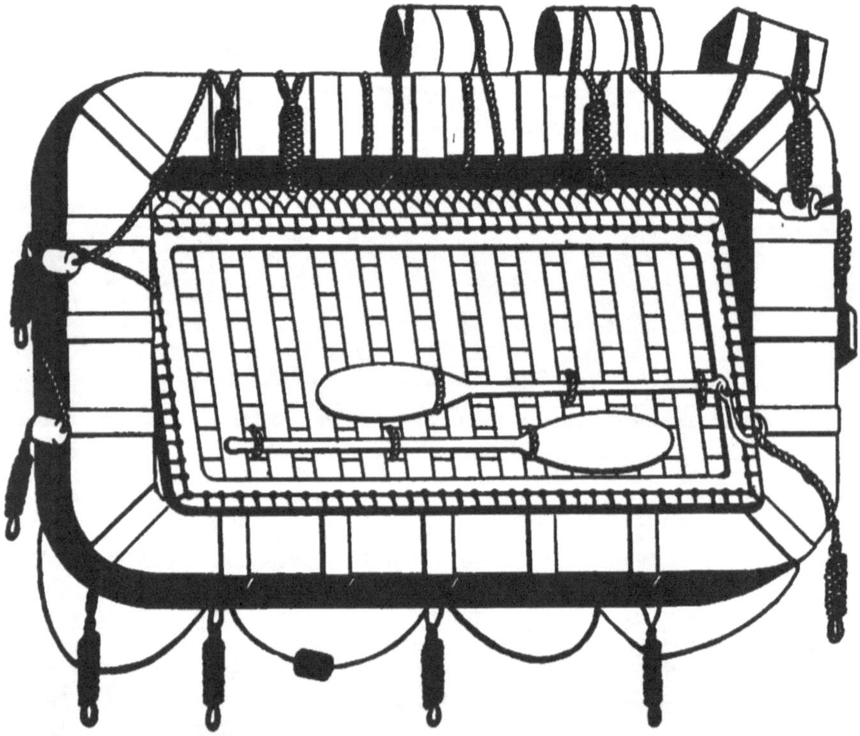

Figure 3. One type of life float.

8. CARBON DIOXIDE RUBBER BOAT. The carbon dioxide rubber boat is a collapsible rubber boat, used chiefly by airmen, inflated by carbon dioxide contained in a cartridge attached to the boat (fig. 4). The capacity of the boat varies, but equipment carried in bags along the inside of the boat usually includes a compass, first-aid kit, fishing kit, two collapsible oars, provisions, sail fabric, signaling mirror, smoke grenades, and a supply of water.

Figure 4. Carbon dioxide rubber boat.

9. LIFE PRESERVERS AND ATTACHMENTS. a. Life jackets. In addition to having a life preserver for each person allowed to be carried, all vessels are provided with an extra 25 percent of this number stowed so as to float free of a sinking vessel. Life jackets are made of buoyant material such as cork, balsa, and kapok, or are inflated by mouth or by carbon dioxide cartridges. Some types are so made that the cotton tape straps may be tied even though the jacket is worn inside out (fig. 5). In wartime, in emergencies, and when either fire or boat drills are being held, the life jacket is worn like a vest. It is not worn when jumping into flame- or oil-covered waters, when swimming under-water to avoid flames or oil, or when jumping from a great height.

(1) *Cork or balsa jacket.* This type consists of blocks of cork or balsa wood sewn into a sleeveless canvas jacket. It is tied about the waist with cotton straps (fig. 5). A cork or balsa jacket should not be worn when jumping from a vessel, even from moderate heights, because it may cause serious injury to the neck or head upon impact with the water.

(2) *Kapok.* Life jackets made from kapok are more comfortable, warmer, and safer to use than cork or balsa jackets. The Army-type kapok jacket is similar to the Navy-type, except that the Navy jacket has an additional kapok collar to support the head. (See fig. 6.)

(3) *Inflation ("Mae West").* The inflation-type ("Mae West") life jacket is made of rubberized fabric and inflated by carbon dioxide cartridges or by mouth. When inflated, carbon dioxide or air fills tubes in the front of the jacket and around the back of the neck. It is worn by flying personnel as it is light in weight and not bulky. (See fig. 7.)

b. Life belts.

(1) *General.* Life belts are made of rubberized fabric in the form of tubes which are inflated by carbon dioxide cartridges or by mouth. They are put on and worn like cartridge belts. Life belts are generally issued to amphibious assault troops, although other preservers may be issued.

Figure 5. Cork (or balsa) life jacket, inside strap tied.

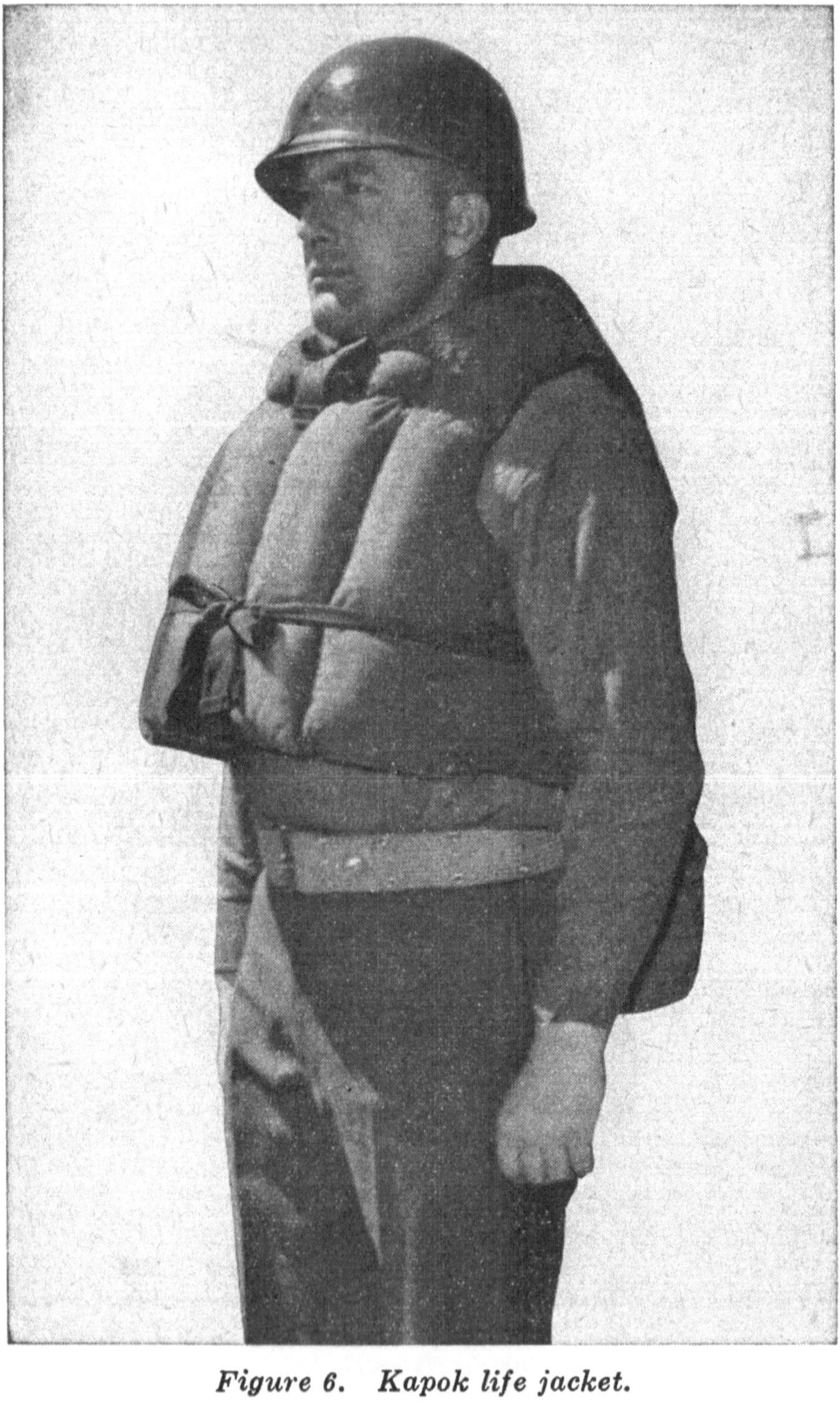

Figure 6. Kapok life jacket.

Figure 7. Inflation-type life jacket being inflated by cartridge.

(2) *Preparation for wearing.*

(*a*) Close valve at end of each mouth-inflation tube.

(*b*) Secure pleats in belt by snapping two fasteners on opposite sides of belt about midway between ends of mouth-inflation tubes.

(*c*) Shorten belt by forming loop and snapping three fasteners (fig. 8).

(*d*) Unscrew caps of inflation mechanism and insert a filled carbon dioxide cartridge into each compartment with slender portion of cartridge pointing toward removable caps (fig. 9). Never reverse the position of cartridges.

(*e*) Replace caps and, using finger pressure only, screw down tightly to prevent loss of gas when belt is inflated.

(3) *Putting on belt.* Grasp inflating mechanism with right hand and put on like a cartridge belt. Hook ends together with slotted tongue on inflating mechanism (fig. 10). Do not turn belt *in* against body. Accidental inflation in such a position makes it impossible to remove without injuring belt. It should fit comfortably about waist. Hook may be adjusted by rotating it to unclamp it, moving it to desired position and reclamping it. Small metal rings are set into belt for attaching improvised rope suspenders, if desired. Suspenders are not furnished with belt.

(4) *Inflating belt.* Grasp belt to right of inflating mechanism, as illustrated in figure 11.

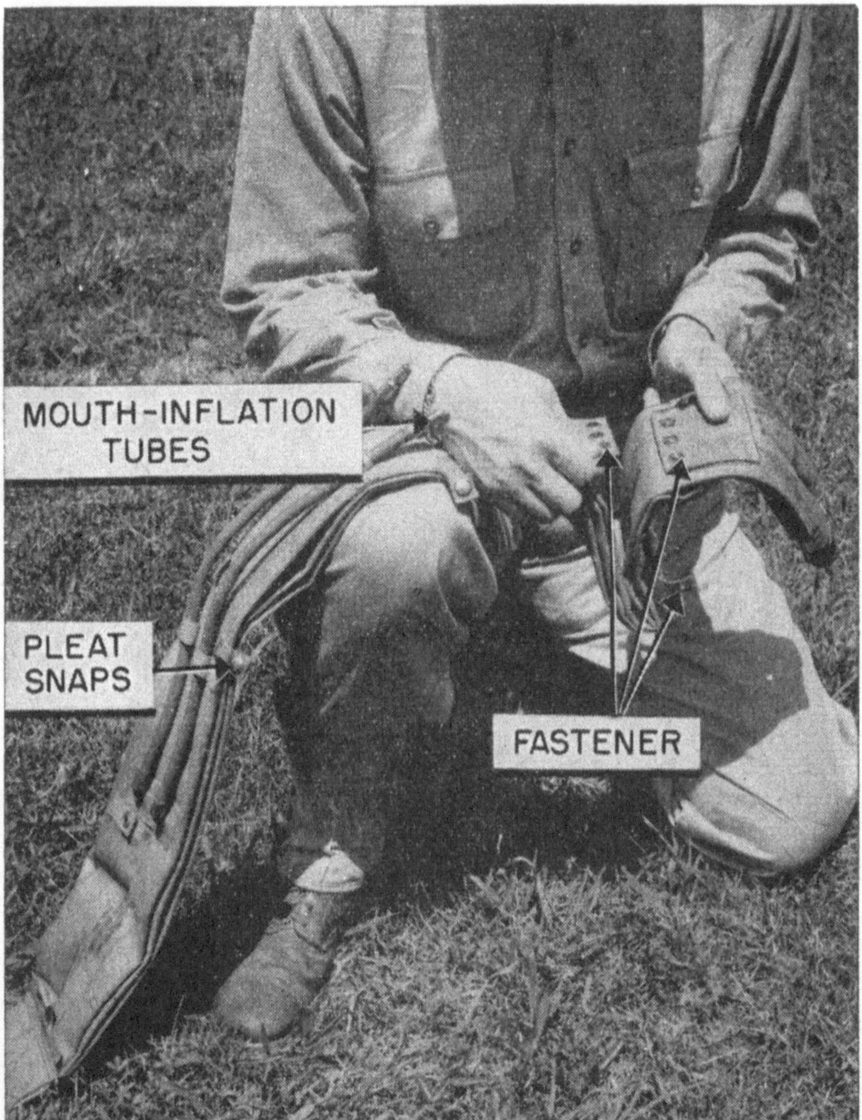

Figure 8. Shortening the life belt.

Close hand quickly and firmly, and then release it. This movement forces a pair of levers to press the carbon dioxide cartridges against firing pins in the removable caps. The cartridge diaphragms are punctured and

Figure 9. Inserting carbon dioxide cartridges into life belt.

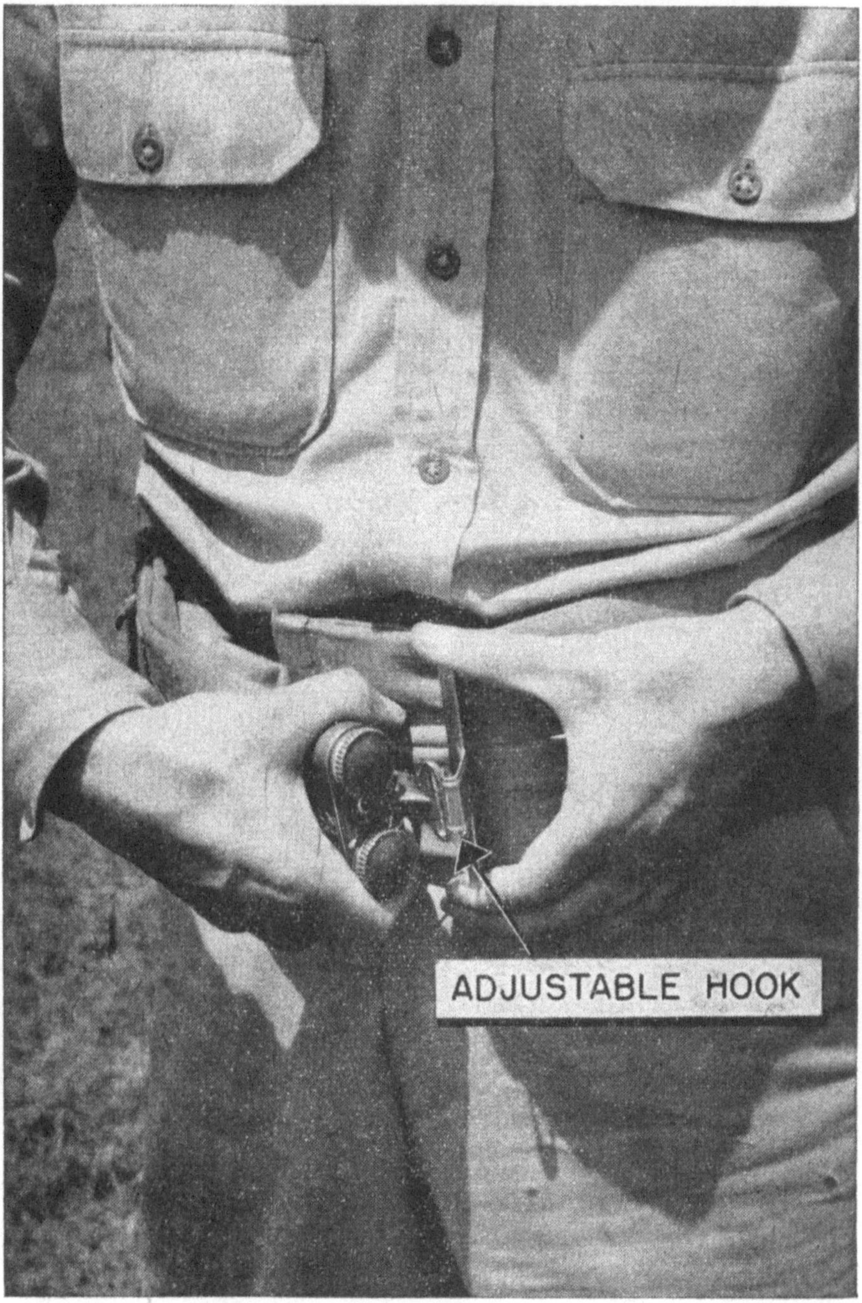

Figure 10. Putting on life belt.

the carbon dioxide is released, inflating the belt. If additional buoyancy is required in either of the buoyancy tubes, reach around with the left hand, pull on the desired mouth-inflation tube (fig. 8), unscrew the valve, and inflate by mouth. The valve must then be screwed tight by hand. To deflate, unscrew caps of inflation mechanisms. In inflating the belt solely by mouth, all snap fasteners must first be disengaged by hand.

(5) *Rules for wearing.* The following general rules apply to the wearing of the carbon dioxide life belt:

(*a*) The life belt should be partially inflated by mouth at all times.

(*b*) It should be worn under personal equipment. The wearer may discard his equipment without removing the belt if, for example, he finds his equipment is too heavy for the belt to support in the water.

(*c*) The life belt should be inflated before going overboard, unless the water is covered with flames or oil. However, inflating life belts before jumping overboard may result in injury to the wearer or damage to the belt.

(*d*) Life belts should normally be worn high on the waist to keep the head above water if the wearer is stunned or injured in leaving the ship.

c. Life rings. Life rings or life buoys are placed aboard a vessel so as to be easily accessible. They are never permanently secured in any way and will float free of a sinking ship.

Figure 11. Inflating life belt by cartridges.

d. Attachments to life preservers.

(1) *Light.* The waterproof life-preserver light, battery-operated, attracts attention to a man in the water. The red light, used to distinguish individual men from rafts and floats (which show white lights), is turned on by screwing on the red cap as far as it will turn. The light has a cord and large safety pin by which it is pinned to the jacket. It also has a clip on its side by which it is secured high on the jacket to gain maximum visibility and keep it out of the water.

(2) *Whistle.* The whistle is worn around the neck. It is valuable in attracting attention at night or in fog. Immersion in salt water for a long time may affect the voice, but it will still be possible to blow a whistle.

10. EXPOSURE SUITS. a. Description. Many vessels under military control, as well as military aircraft operating in cold latitudes, carry an exposure or lifesaving suit for every member aboard. Worn with life jackets, the suits can be donned quickly and provide floatability, water-proof comfort, and retention of body heat. Several layers of loose woolen clothing may be worn inside the suit, and the wearer can rest after being set adrift by raising his legs and floating horizontally.

b. Precautions. Exposure suits usually do not prevent water from entering about the neck, but if a towel is wrapped around the neck when the suit is worn, less water will enter. Should the suit be punctured and fill with water, the life preserver, worn on the outside, will continue to keep the wearer afloat.

Most exposure suits are made from rubber and should be kept away from heat, acids, grease, oil, and abrasive surfaces. Repair kits are furnished, along with talcum powder which helps preserve the rubber.

11. LINE-THROWING GUN AND BREECHES BUOY.

a. Beach patrol. Personnel at Coast Guard stations patrol the nearby beaches between sunset and sunrise for about 4 miles on each side of the station. In wartime or when the weather is foggy, the patrol is maintained day and night and a continuous watch is kept at the station itself. When a vessel is immobilized, regulation distress signals are used to attract the attention of those ashore or on nearby vessels. (See par. 59.)

b. Line-throwing guns.

(1) *General.* The line-throwing gun is standard equipment on all oceangoing vessels and is operated and manned by the ship's personnel. These guns and breeches buoys are standard equipment at all Coast Guard life-saving stations and aboard all Coast Guard vessels. Line-throwing guns are used to carry a small line to shore or to another vessel from a ship in distress, by projecting a shank to which the line is attached. The gun may be lashed to the deck and fired at a high elevation to shoot over and across the target. Shoulder guns (fig. 12) are also used.

(2) *Equipment aboard ship.* Equipment for the line-throwing gun, except powder and primer, is stowed on the boat deck in a box marked "Line-Throwing Gun." Such equipment consists of six projectiles (each projec-

Figure 12. Line-throwing shoulder gun.

tile is a shank with an eye); four shot lines, each 1,700 feet long; and an auxiliary line, 3 inches in circumference and 1,500 feet long. Powder (in 2- and 5-oz charges) and primers are usually stored in the ammunition locker or in the master's room.

(3) *Firing.* Experienced ship's personnel will supervise the firing of the line-throwing gun. Instructions are usually provided with the equipment.

c. Rigging and operating the breeches buoy. The breeches buoy consists of a ring buoy with a canvas, trouserlike bottom in which a person being rescued is conveyed to safety by ropes secured in clothesline fashion (A in fig. 13). After the line shot from the line-throwing gun is received, additional lines and tail-blocks are hauled along this line until a pulley line can be secured between the rescue party and those in distress. The breeches buoy is attached to this line and put into operation to carry persons to safety. B in figure 13 illustrates details of the rigging of a breeches buoy.

Section II. SAFETY MEASURES

12. GENERAL. a. Assignments. Abandon-ship stations are assigned by the master to all personnel aboard. Routes to stations are explained and such directional signs as arrows are pointed out. Lifeboats or rafts are also assigned. Boats are numbered from forward towards the stern, with odd-numbered boats on the right (starboard) side and even-numbered boats on the left (port) side. If boats are nested, No. 1–A is carried under No. 1, No. 2–A under No. 2, and so on. The number and location of the boat assigned must be known thoroughly, so that it may be found even in the dark.

b. Inspection. An inspection of the troop compartments is made daily for cleanliness and a count of personnel.

c. Security. Security measures during peacetime will vary from those prescribed in time of war. However, these security measures will be prescribed by the commander concerned.

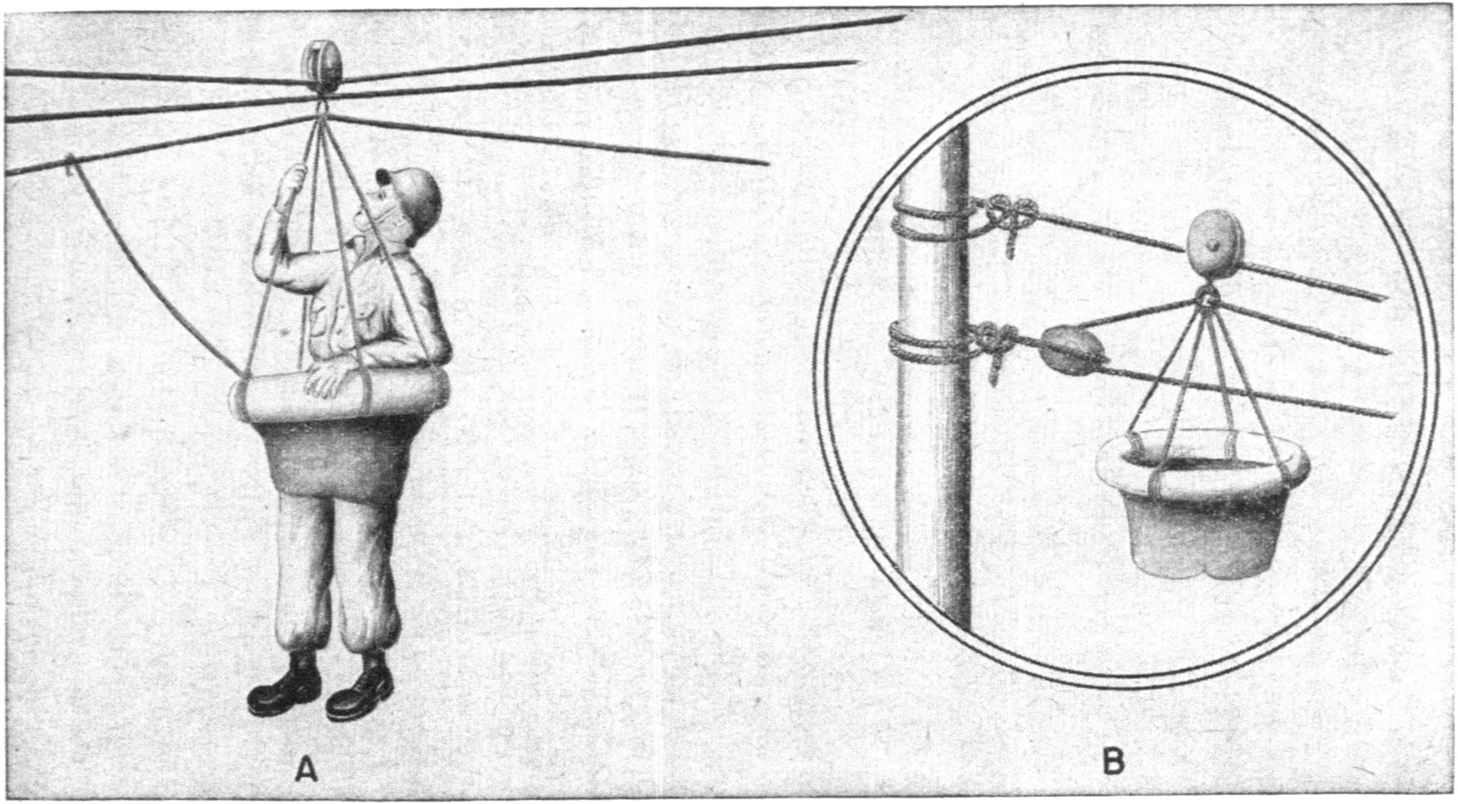

Figure 13. Breeches buoy.

d. Abandon-ship drills. These drills are usually held without warning and simulate real emergencies as much as possible. In peacetime, Coast Guard inspectors usually conduct a presailing drill. At sea, drills are first conducted by day to enable passengers to become familiar with locations of ladders, cables, equipment, and projecting parts so they can find their way quickly to stations in the dark. Alternate routes to stations are prescribed. At different times, one or more routes are blocked off so that passengers must use others and thus become familiar with them. Instruction in leaving the ship during an emergency includes use of safety equipment, special precautions, and conduct in the water and on lifeboats and life rafts.

13. INDIVIDUAL SAFETY PROCEDURE DURING PASSAGE. a. General precautions. Although over-all precautions will vary with different ships and situations during times of emergency, the following rules should serve as a guide to all personnel on Army ships in both peacetime and wartime. Every passenger should—

(1) Smoke only at times and places designated.

(2) Keep out of the way of the ship's crew while they are at work. Traffic on deck moves counterclockwise: on starboard—toward the bow; on port—toward the stern.

(3) Learn thoroughly the procedure for meeting such emergencies as abandoning ship, fire, and attack.

(4) Know the emergency signals for all ship's drills, and know the location and duties of

his emergency station. Prompt obedience to all emergency signals is absolutely essential.

Signals

FIRE AND EMERGENCY ..	Rapid ringing of the ship's bell and continuous ringing of general alarm bells for a period of at least 10 seconds.
ABANDON SHIP..	More than six short blasts and one long blast on the whistle and the same signal on the general alarm bells.
MAN OVERBOARD	Shout "Man overboard" to the bridge.
DISMISSAL	From fire and emergency stations, three short blasts on the whistle and three short rings on the general alarm bells.

Whistle Signals Used for Handling Boats

LOWER BOATS...	One short blast on the whistle.
STOP LOWERING BOATS	Two short blasts on the whistle.
DISMISSAL FROM BOAT STATIONS	Three short blasts on the whistle.
SWING OUT BOATS	Four short blasts on the whistle.
MAN EMERGENCY BOAT ..	One long and one short blast on the whistle.

(5) Keep fit, get all the exercise he can, and apply himself vigorously in any planned exercise or athletics.

b. Additional wartime precautions. In wartime, the following rules should be observed in addition to the general precautions discussed in *a* above:

(1) Always have a life jacket or life belt, whistle, rope, life-preserver light, cartridge belt, and a full canteen of water present. Wear the jacket or life belt during the day; when not worn (e.g. when retiring) it is kept within reach.

(2) Practice finding essential personal equipment in the dark. Sleep as nearly fully clothed as conditions permit.

(3) Always dress warmly. Uniform to be worn will be specified by regulations. Adequate clothing offers protection from flash burns and other hazards during an emergency at sea.

(4) If on an exposed deck when bombs hit the ship, lie flat on the deck to avoid blast and flying fragments. If it is impossible to lie down, hold on to something solid to avoid being thrown.

c. Precautions for cold climates. When sailing through Arctic waters, wearing of the prescribed clothes, including gloves and face mask (on lookout duty), is especially important. The following factors should also be noted:

(1) Several layers of light clothing are much warmer than a single heavy layer. Inner garments should be fluffy and porous, outer garments rainproof and windproof.

(2) Clothes should be loose enough to permit air circulation so that evaporation of perspiration can take place. Otherwise frost will form inside clothing with harmful results. Shoes should be big enough to allow the wearing of two or more pairs of socks.

(3) Try to keep clothes dry, especially socks. If the weather is very cold, keep clothing on, wet or not. Dry it out as soon as possible.

(4) Wear heavy woolen inner gloves protected with windproof and rainproof outer mitts.

(5) Don't become overheated. Wear clothes which can be removed in layers.

(6) Brush frost off clothes before going in warm places. Melted frost causes clothes to become damp.

(7) Clothes should be kept clean, as oils from the skin fill the air cells in the cloth, thus reducing their insulating qualities.

(8) Don't grease boots as grease is a poor insulator and may freeze.

(9) Wear watch on string around neck and next to body. If it becomes too cold, it may stop.

(10) Don't touch cold metal with the bare skin; it may freeze fast to the metal. If this should happen, don't try to pull loose. Thaw the metal.

(11) Wear dark glasses, if possible, to protect the eyes from sun and snow glare.

(12) Beware of slippery decks and ladders covered with frozen spray.

(13) When an Arctic shore is reached after abandoning a ship, knowing what to do may save a person's life and bring rescue. (See par. 61.) To survive in a cold country, remain calm and avoid overexertion; eat plenty of fat, if possible; avoid tight clothes and shoes; and keep as warm and as dry as possible. A knowledge of appropriate signals

will be helpful in attracting the attention of rescue parties (pars. 52–56).

d. Precautions in tropical waters.

(1) Rapid cooling of the body after excessive sweating should be avoided because it reduces body resistance which may result in pneumonia, bronchitis, cramps, or skin conditions (prickly heat and fungus).

(2) Change wet clothing as soon as possible.

(3) Don't wear tight clothes that restrict air circulation.

(4) Drink plenty of water.

(5) Take daily doses of salt tablets to replace body salt lost through sweating.

(6) Physical cleanliness is essential. Try to keep bowel movements regular.

(7) Wear sun glasses for protection from water glare.

(8) Be careful to avoid sunburn. The tropical sun can burn a person seriously before he realizes it.

(9) Know how to recognize and deal with heat stroke, heat exhaustion, and heat cramps.

(10) Learn methods of survival in the jungle. (Refer to par. 61.) A person can survive if he keeps calm and reasonable. Good food and pure water are available if looked for. Most wild animals, including poisonous snakes, will not attack human beings unless they are first bothered or annoyed. The majority of the jungle natives are friendly if friendly gestures are made toward them, but don't be overbearing. The worst enemy is malaria, but it can be avoided by taking

quinine, atabrine, or other antimalarial drugs. In addition, know how to make the correct signals to attract attention for rescue (pars. 52–56).

CHAPTER 3

ABANDONING SHIP

Section I. ABANDON-SHIP CONDUCT

14. REASONS FOR ABANDONING SHIP. a. General. Most casualties at sea are actually the result of panic, which is usually the product of fear based on misinformation and a lack of self-confidence, training, and knowledge of proper actions in an emergency. Routine drills in abandoning ship, instructions in sea survival procedures, explanations of reasons for abandoning ship, and emphasis by good leadership on the chance of survival for those set adrift will keep passengers and crew alert and prepared for any emergency.

b. Natural hazards. Natural hazards at sea include:

(1) *Ice.* Icebergs usually have a mass only one-eighth to one-twelfth visible on the surface of the sea. These masses of ice are found north of latitude 40° N and south of latitude 35° S. In addition to the danger of ramming icebergs in these areas, ships may be stranded and smashed by forceful ice flows.

(2) *Fire.* Fire is a hazard always present and is one of the most terrifying experiences aboard ship. Personal care with cigarettes and matches reduces the danger.

(3) *Other causes.* In crowded harbors, there is danger of collision or running aground, espe-

cially in fog. Mechanical failures, such as a damaged rudder, may put the vessel at the mercy of the sea.

c. Hazards from enemy action. Natural hazards are intensified by such wartime restrictions placed on navigation as radio silence and blackout. Additional dangers from enemy action include torpedoing, bombing, explosion of depth charges, mines, shell fire, and sabotage.

(1) *Torpedoing.* Torpedoing is the principal menace at sea during wartime and may be caused by underwater, surface, or air attack. In spite of the hazard, a large percentage of men involved in torpedoings have survived without permanent ill effects. There is no time at sea when vigilance can be relaxed. Twilight, dawn, and changes of watch have been the times when attacks have been most frequent.

(2) *Bombing and depth charges.* Moving ships are small targets for bombing planes, but bombing is a major wartime menace. Dive bombing is often effective, and considerable damage can be caused by strafing. A person within a 30-foot radius of a bursting bomb will generally not survive the concussion, but the chief danger lies in the thousands of steel fragments thrown from the casing of a bursting bomb. These fragments travel as far as a mile, but beyond 200 to 400 feet, depending on the size of the bomb, fragments are not likely to be effective. A depth charge exploding within 30 feet from a ship's side can push in the hull plates.

(3) *Mines.* Mines may be sowed by underwater craft, surface craft, or planes, and may be found singly, in regularly patterned fields, or in irregular patches. Free-drifting mines can be considered the worst hazards. Individual mines may be camouflaged by being hidden under boxes, wreckage, or floating debris.

(4) *Shelling.* Danger of shelling may exist when a ship is within range of enemy shore batteries, planes, surface vessels, or surfaced submarines.

(5) *Sabotage.* Sabotage by enemy agents aboard ship may be considered an additional danger.

d. Chances for survival. Half the battle is won when an individual reaches his abandon-ship station. Chances of survival are excellent for those with foresight, knowledge, training, good physical condition, and initiative.

15. PROCEDURE WHEN ABANDON SHIP SIGNAL IS GIVEN. **a. Proceed to abandon-ship station.** Don't become excited when the signal for abandon ship (par. 13*a*(4)) is sounded. When the alarm is heard, proceed rapidly and calmly to the assigned abandon-ship station. *Do not run.* Casualties caused by panicked persons outnumber those caused by any other means. The most important factor in survival at sea is being prepared when the order to abandon ship comes. Even if all mechanical communications have broken down, word-of-mouth communication always may be used. If individual equipment is lost or misplaced, go to the station without it. Many men have lost their lives and endangered the lives of others by returning to their quarters to get something they forgot. Keep silent so that orders can be heard. Before the signal to abandon ship is sounded, all persons should be given the following information: The ship's approximate location, the direction and distance to the nearest land, and the result of SOS signals. If a rescue ship has answered the SOS, lifeboats remain near the spot where the ship was abandoned. Otherwise, the location of the ship and distance and direction to land can be used to steer a course.

b. Wait for orders. Do not get into the lifeboat or attempt to lower it. Wait for orders. Normally, certain crew members are in charge of each lifeboat. If one of the crew assigned is a casualty, be prepared to assist in lowering the boat. Otherwise, keep out of the crew's way. If the lifeboat or raft is damaged, do not get into another without permission of the man in charge of the lifeboat to which originally assigned. Do not risk injuring occupants by throwing anything down into a lifeboat.

Section II. GOING OVER THE SIDE

16. GENERAL. a. Precautions. In abandoning ship, leave by the method assigned (line, cargo net, ladder, lifeboat, etc.). All persons act according to orders, even in extreme emergency. If the order "Every man for himself" is given, get into the water and at least 50 yards away from the ship as soon as possible. Before getting into the water, look carefully about to avoid flaming oil or floating debris. Jump only when it is impossible to go down a hose, line, net, or ladder. and before using a hose or rope, make sure it is secured to the railing or a stanchion. Act quickly, but not rashly. After entering the water, try to reach a boat, raft, or any other object that will furnish support.

b. Jumping from ship. If there are no other means of going overboard, then jump, leaving the ship by the bow or stern, whichever is lower in the water. Always remember that the higher the point from which the jump is made, the more chance there is of being injured. If possible, always jump into the wind and swim away from the ship in that direction; this will put the greatest distance between survivors and the ship in the shortest possible time.

17. LIFEBOATS. a. General. A lifeboat commander and crew made up from the vessel's crew are assigned each lifeboat. This group takes charge of the passengers and lowers the lifeboat into the water. In the absence of such a crew or when only one crew member has been assigned to the lifeboat, anyone may be called upon to assist in lowering and launching the lifeboat.

b. Suspension of boats. Lifeboats are suspended on two metal members called davits. The lifeboat is fastened to the davits by two pairs of blocks, one pair at the bow and one pair at the stern of each lifeboat. The line running from each upper block is called the boat's fall. Paying out the fall lowers the boat. Davits may be one of three types: gravity (fig. 14①), radial (fig. 14②), or quadrantal (fig. 14③).

c. Boarding the lifeboat. Passengers usually board a lifeboat from the main (passenger) deck where the boat is held before it is lowered into the water. Under certain conditions, the coat may be completely waterborne before it is loaded. A man should make sure that his life jacket is properly adjusted when he boards the lifeboat.

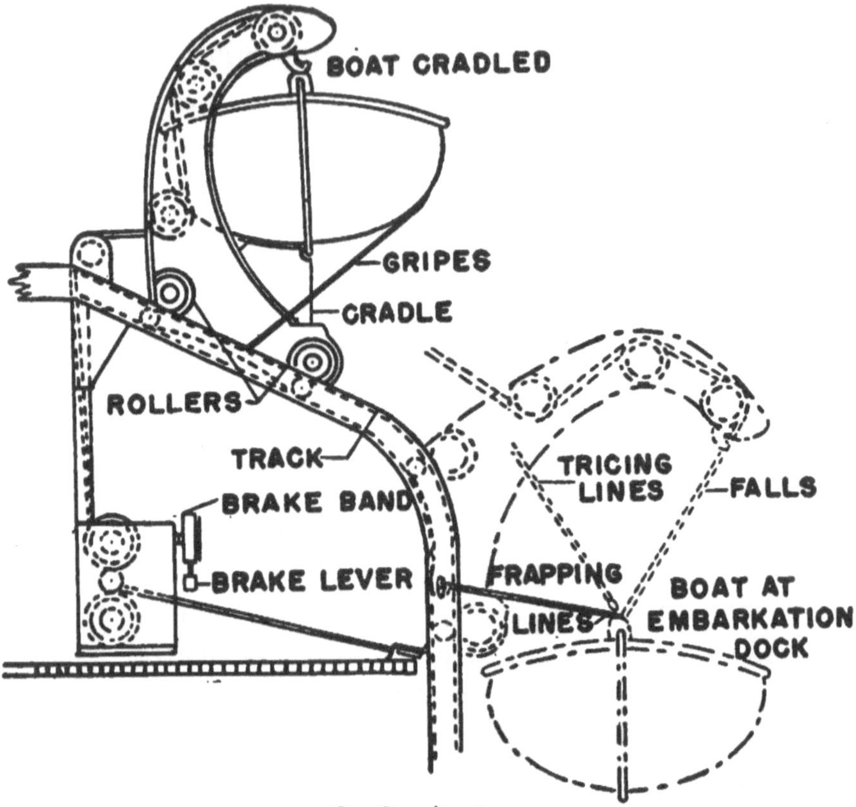

① Gravity type.

Figure 14. Types of davits.

18. NETS. a. General. Life nets or cargo nets are normally hung over the side of a ship and are used as broad ladders for disembarking into landing craft or abandoning ship.

b. Descending nets. On troopships or similar large vessels with sufficient freeboard (the part of the hull above the waterline), the men start over the ship's side in groups of four to six at a time. When the first line has descended halfway, another line starts over the side. Thus there is always a line starting overside, a line halfway down, and a line at the lower end of the net. In the case of Liberty ships and other vessels of lesser freeboard some modification of this

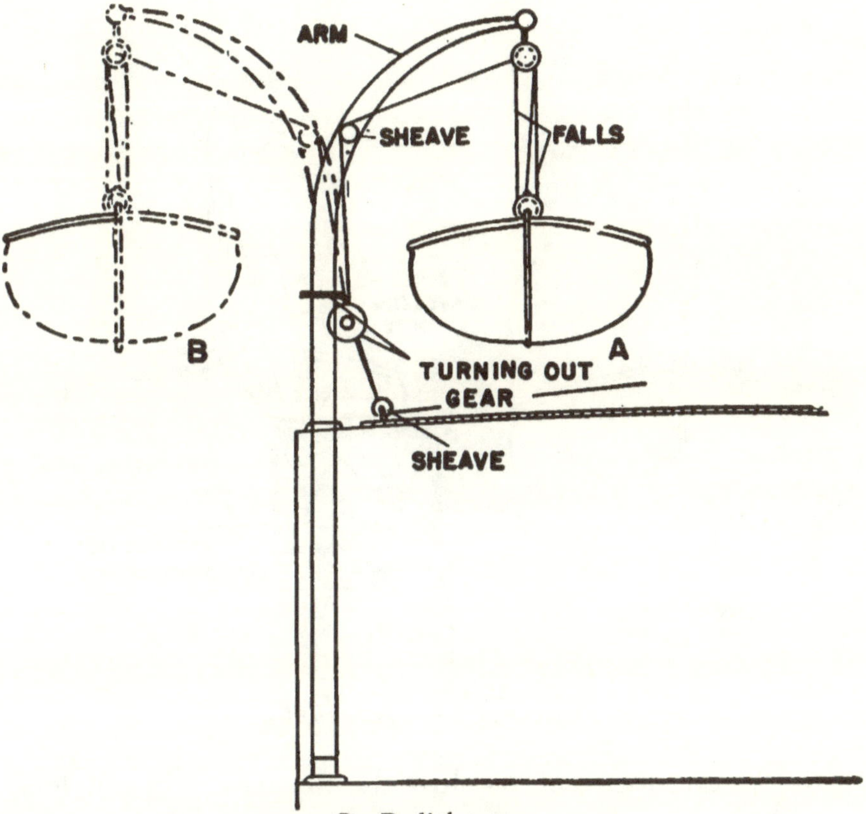

② Radial type.
Figure 14—Continued.

method should be employed. There are two methods of climbing down life nets.

(1) Grasp a single vertical strand, and place the feet on the horizontal strands on either side of the vertical strand (A in fig. 15).

(2) Grasp the outside strands of a group of three vertical strands, and place the feet on either side of the center strand (B in fig. 15).

(3) In either method, grasp vertical, not horizontal, strands of the net so that the hands will not be stepped on by a man descending above. Keep the hands well above the head, and move the feet down one square at a time,

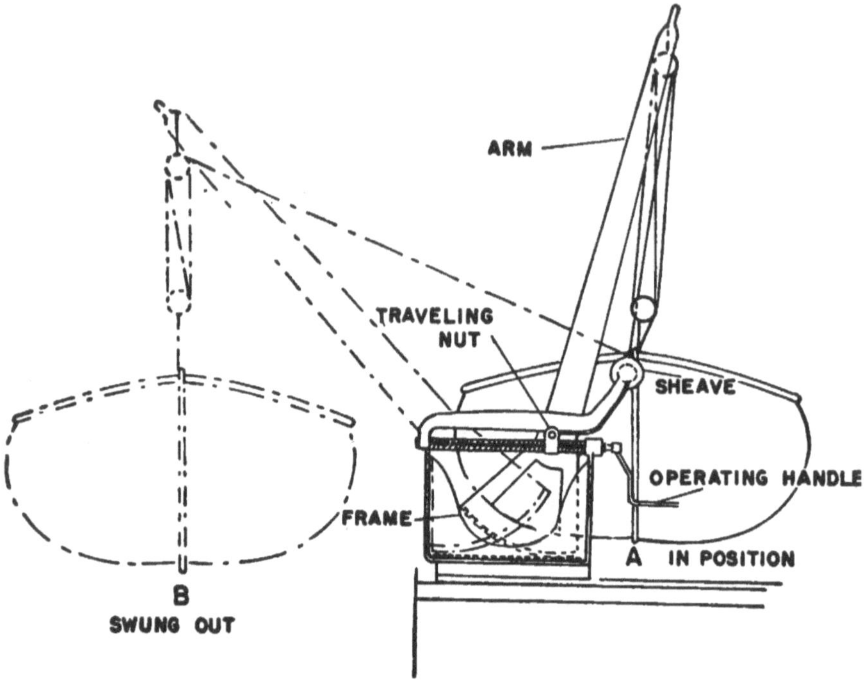

③ Quadrantal type.
Figure 14—Continued.

as longer steps slow the descent. Do not look up or down, but keep the head level, as falling objects from above may result in serious facial injuries if the descent is made with upraised face.

c. Dropping from the net. On vessels not heavily loaded, nets may not reach the water and it may be necessary to drop from the end of the net into the water. To drop into the water, place both feet on the same horizontal strand near the bottom of the net. Bring the hands down until the body is in a crouched position with arms bent. Kick your legs back and push with your arms to bring your body into a vertical position as you drop into the water.

d. Lowering survivors. To lower a disabled survivor over the side, place the man on a pallet,

Figure 15. Descending a life net.

stretcher, or other firm horizontal surface. Place the pallet or stretcher in the middle of an unattached cargo net, gather the lengthwise net corners together, and attach a lowering line. By this means, the disabled survivor can be lowered into a lifeboat or raft with the least possible risk of further injury.

19. LADDERS. a. Precautions. Ladders must be descended slowly. Be careful that the hands are not mashed against the side of the ship when the ladder swings in and out.

b. Jacob's ladder. This is a rope ladder with flat steps. Grasp the vertical ropes and move your feet down one step at a time. The hand and leg on the same side of your body should move at the same time.

20. ROPES. a. General. Ropes can be secured to parts of a ship, thrown over the side, and used to descend into the water. It is always better to go over the side on a rope than to jump. Use gloves if available. Don't slide, as sliding may burn the hands seriously.

b. Hand grip. Climb over the side and grasp the rope with the feet before grasping with the hands. Descend slowly, hand-under-hand, bending the arms slightly. There should be a sensation of holding and gripping, rather than hanging.

c. Leg grips. Use of a leg grip depends on the tautness or slackness of the rope, its size and weight, the length of descent, and the condition of the hands.

(1) *Taut rope.* On a taut rope, cross the legs with one knee drawn up and the toes lifted. The rope runs along the inside of this leg, over the front of the ankle, and down the

outside edge of the foot. Cross the other foot over to hold the rope between the outside edges of the feet near the heels. Pressure with knees and feet slows the descent (fig. 16).

(2) *Loose rope.* When the rope is sufficiently loose, either of two leg grips may be used.

(*a*) *Stirrup grip.* Hold the legs straight and together so that the rope lies along the outside of one leg, under the foot, and over the foot of the other leg (① in fig. 17). Pressure of one foot against the other regulates the speed of descent. The hand-under-hand method may be used, or the hands may slide together, taking a firm hold when foot pressure stops the descent. If the hands are injured, hug the rope with the arms.

(*b*) *Secure-foot grip.* This grip is best for long descents when the arms may need to be rested. Allow the rope to drop between the legs and across the instep of one foot. Step on the rope with the other foot where the rope crosses the instep, and by applying pressure, grasp or release it (② in fig. 17).

d. Knotted ropes. Ropes used for abandoning ship usually have knots at frequent intervals, approximately 18 inches apart. These knots can be used to allow muscular relaxation and to prevent slips and rope burns. Grasp the rope with the hands just above a knot. Keep the feet together, press the rope between the insteps, and bend arms and legs slightly. Release the feet and lower them to the next knot. Hold on

Figure 16. Descending taut rope, hand-under-hand.

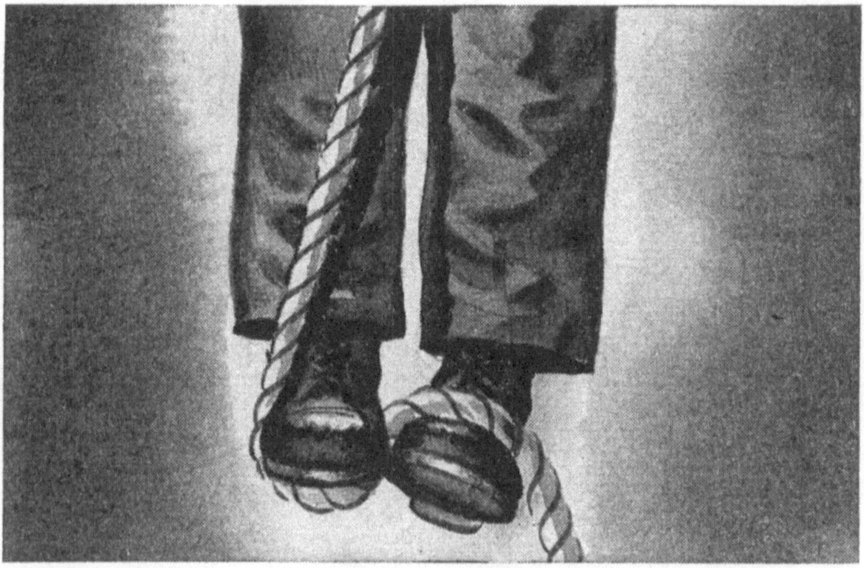

① Stirrup grip.

Figure 17. Leg grips for descending loose ropes.

② Secure-foot grip.

Figure 17—Continued.

with the feet and bring the hands down to the next knot.

e. Deliberate descents. Rope descents should be

deliberate and unhurried, legs and feet applying pressure to prevent arm strain and consequent slipping. Long or fast drops or slipping will cause severe rope burns.

f. Discipline. When a single rope is being used by several men, they should keep well separated. Look out for men coming down, because it is possible to slip and fall on others below.

g. Climbing a rope. Grasp the rope with the hands as far up as possible and, holding with the hands, bring up the legs as far as possible. Grip the rope with the legs, using the taut rope leg gr p as illustrated in figure 16. Relax the hands and reach for a new grip while holding on with the legs. Move the legs up for a new grip while holding on with the hands.

21. JUMPING. It is safer to jump than to risk a head injury by diving into water of unknown depth or into swimmers, lifesaving craft, or debris. Further, while jumping, it is also easier to observe the surface of the water up to the moment of impact.

a. Procedure. Remove helmet before jumping. With the left hand, pinch the nostrils together and with the right hand clutch the left shoulder, or vice versa. The arm across the chest protects the face from impact with the water. Hold the head erect, body relaxed and straight, and legs together.

b. Precautions.

(1) *When there is no oil on the water.*

(*a*) Take a deep breath before jumping.

(*b*) With the downward roll of the ship, step forward as though taking the next stride, and, springing from the other foot, bring the legs together in the air. Drop vertically into the water. Swim quickly out of the danger area.

(*c*) Do not look down. Keep the head erect; chin in.

(*d*) Jump feet first as far from the ship as possible.

(*e*) Do not try to break the fall with hands or arms.

(*f*) Keep legs together. After entering the water, open them to check the depth of the plunge.

(*g*) Do not wear a life jacket if jumping from a great height. The life jacket should be held in one hand, and the other hand should be used to hold the nose and protect the face. Hold the life jacket by the tape, and if it is jerked from the hand by the impact with the water, recover it as soon as possible. The life jacket may also be held with a short piece of rope and both arms and hands used to protect the nose, face, and chest.

(2) *Jumping into oil or flames.* Ships usually carry their fuel oil in tanks on either side of the ship. The explosion of bombs or torpedoes may burst the tanks, releasing the oil and allowing it to spread on the surface of the water. Oils are classified as thin oils and thick oils. Fuel oil for ships is a heavy oil of thick consistency and difficult to move through. It may form layers on the surface of the water several inches thick. Whenever possible, avoid jumping into water cov-

ered with fuel oil, and do not try to swim through it. Although fuel oil is only slightly inflammable, the following precautions should be observed when it is necessary to jump from a ship and there is flame or oil on the water.

(*a*) Remove life preserver and anything else that might carry you to the surface into the oil or flame. Remove shoes, but wear shirt, trousers, and socks.

(*b*) To prevent trapping air under the clothing, fasten all buttons on shirt and trousers, and tuck trousers into the socks.

(*c*) Close eyes and mouth before entering the water, and hold the nose.

(*d*) If necessary to jump into oil or flame, jump to windward and swim to windward under water, as long as possible. The wind will tend to blow the oil or flame away from you, instead of driving them with you. If it is necessary to come up for air before swimming clear of the oil- or flame-covered area, come to the surface with arms milling violently to cause a disturbance on the surface of the water. This will push flames or oil away from your face and allow you to inhale and continue under water.

22. OTHER MEASURES. a. Shallow diving. The shallow drive is useful in shallow water or water of unknown depth or when it is necessary to cross a body of water quickly, as when escaping from an enemy shore. From the shallow dive it is easy to go into any strong swimming stroke. The dive consists of diving forward when slightly above the surface of the water and arching the body by raising the head and chest and flinging the arms upward. When the body hits the water it skims along the surface instead of going under.

b. Rubber boats. The parachute raft or rubber boat used by airmen is attached to the D-ring of the life jacket by a light webbing. To prepare the rubber boat for use, unsnap the cover, and, by following the retaining line to the inflation valve, disengage the locking device on the carbon dioxide cylinder, and actuate the inflation handle. The raft will force itself out of the case. Be careful to find a clear place in the water before throwing the boat overboard. Rubber boats are sometimes part of a ship's safety equipment; as such, they are discussed in paragraph 8 and illustrated in figure 4.

c. Fire hose and boat falls. Fire hoses are usually about 50 feet long. They can be made fast to the deck, let over the side, and used in the same manner as ropes for leaving the ship. Lifeboat falls may be similarly used after the lifeboats have been lowered.

Section III. IN THE WATER

23. PRECAUTIONS. a. Getting away from the ship.

(1) *In lifesaving craft.* Lifesaving craft should be moved rapidly at least 50 yards from the ship to escape the suction usually caused when the ship sinks. A slowly sinking vessel, especially one going down by the head

or by the stern, may submerge without creating a suction.

(2) *Swimming.* Once in the water, immediately move away from the ship. The danger of injury from underwater explosion is lessened by swimming or floating on the back. Swim or paddle to a lifesaving craft or any other floating object that will furnish support. If this is not possible, move at least 50 yards from the ship. When beyond this danger zone, remember that buoyance is the main consideration. Unless land is in sight the distance you swim is relatively unimportant. Retain clothing and shoes as protection from the weather, salt, and oil.

b. Picking up survivors. Up to the point that a boatload is endangered, as many survivors as possible should be picked up. Men in the water may hang on to the life-line around a lifesaving craft. If there is room, men in rafts should shift over to lifeboats. Boats and rafts should stay together to increase the chance of rescue and bolster morale.

c. Rigging mast and sails. Mast and sails should be rigged only when the men have quieted down because, in their excitement, they may fall overboard.

d. Righting capsized boat. If a lifeboat capsizes, five or six men can right it by using the method illustrated in figure 18.

e. Enemy strafing. Enemy aircraft may strafe the boat, but because of its speed, an airplane's attack is brief. Bullets from low-flying planes usually ricochet off the water, and if they penetrate, they will go no deeper than 24 inches. Men in a lifeboat who are physically able to swim should go over the side

Figure 18. Righting capsized boat.

when an enemy plane approaches, and swim or bob 24 inches under water, or at right angles to the plane's line of flight. If sails on the boat are set, they should be lowered to prevent the boat from sailing away while the men are in the water. All men who cannot get in the water should drop to the bottom of the boat. If all go overboard, the strongest swimmers should grasp ropes from the boat to keep it from drifting away. Immediately after the attack, any bullets holes should be plugged with wooden plugs, cloth, or other material.

24. EMERGENCY FLOTATION. The following methods of gaining additional flotation enable a person to remain afloat longer and should be employed by strong and weak swimmers alike.

a. Shirt as support.

(1) To use a shirt as support in the water, button it completely while still aboard ship. Draw the front out of the trousers, hold it down

and forward with arms fully extended 6 to 10 inches from the body, and jump. The shirt will be filled with air when you hit the water and will aid in bringing you back to the surface. By assuming the breast-stroke position on the water, an air pocket is formed in the back of the shirt. This provides some support for a short time.

(2) If you are already in the water and wish to use your shirt for flotation, an air pocket may be formed without removing the shirt. Fasten all buttons, including those of collar and cuffs. Take a deep breath and assume a jellyfish float (par. 26*a*(2)). With the fingers, form an opening in the shirt front between the second and third button, and exhale air into the shirt. The action may be repeated to increase flotation if not enough air is trapped the first time. If there is a sufficient quantity of air in the shirt and a prone position is assumed, an air pocket will form at the back of the shirt and will support the body.

b. Trousers as support.

(1) Remove and wet or dampen your trousers. Tie a knot at or near the end of each leg. Button the fly. Grasp the trousers by the waist, legs down, and hold in front of your body (A in fig. 19). Flip the trousers over and behind the head, arms extended, wrists flexed so that the backs of the hands are down (B in fig. 19). Jump, and as your feet hit the water, snap the hands forward from the wrists to get the waist of the trousers

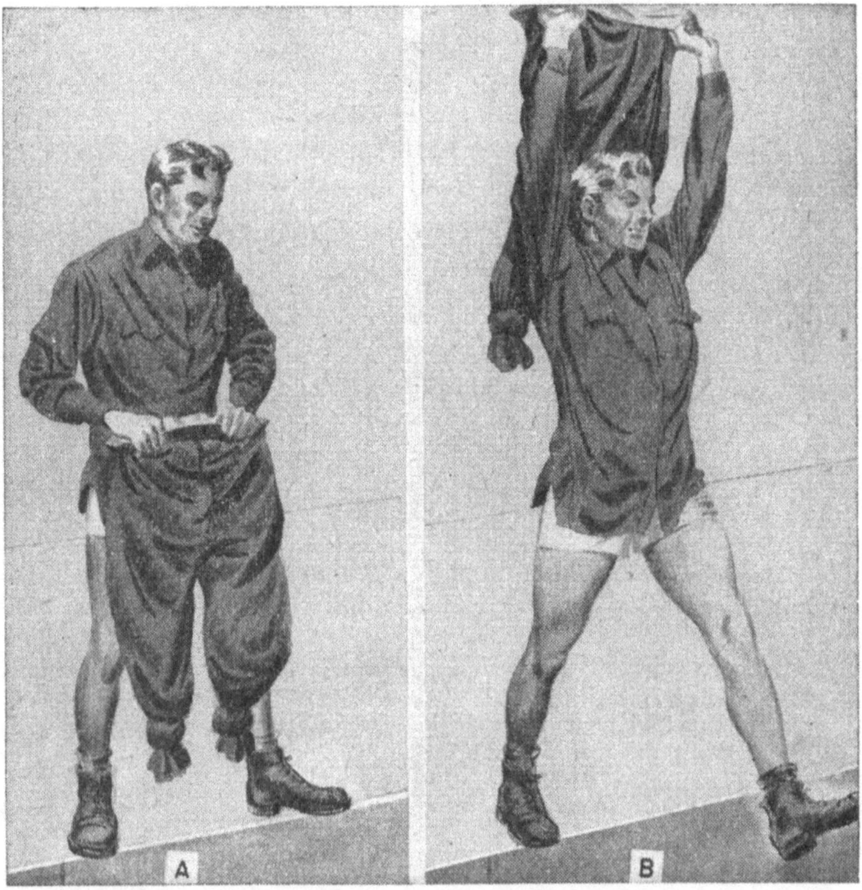

Figure 19. Use of trousers to provide support in the water.

under water (C in fig. 19). The air which is trapped in the legs helps return you quickly to the surface. For surface support, take a prone position and place a leg of the trousers on either side of your body below the armpits (D in fig. 19).

(2) A similar procedure may be used after reaching water. Remove the trousers, tie a knot at or near the end of each leg, and button the fly. While treading water, hold the

Figure 19—Continued.

trousers above the water by inserting one arm in each leg, as illustrated in (a) in figure 20. This allows air to fill each leg. Drop the arms quickly, pulling the waist band under the surface. (See (b) in fig. 20.) This traps air in each leg. The support can then be used in the prone position by placing one trouser leg on each side of the body under the armpits. (See (c) in fig. 20.) If there is not enough air in the trousers, take a deep breath, submerge and hold the waist band below the surface, and expel air into the trousers.

c. Barracks bag or pillow case. Wet or dampen the bag and proceed as with trousers. After entering the water, hold the bag with both arms to prevent its overturning. If sufficient air has not been trapped, take a deep breath, submerge, exhale into the mouth of the bag, and rise to the surface.

d. Sheet, poncho, or squares of canvas. Gather or knot the four corners to form a bag. Proceed as previously outlined.

e. Debris. Any floating debris and wreckage may be used, shared with the greatest number of men. It is better to cling to planks, boxes, and other floating articles than to climb upon them. Clinging to floating debris adds the buoyance of the object to the buoyancy of your body. Lash yourself to the debris if possible. Trying to climb up on an object in the water is difficult and often leads to frustration and rapid exhaustion. Only objects large enough for full support should be boarded. Resting the hands or elbows on an object or throwing the arms around it may provide sufficient support. A plank can be used as a surfboard by lying on it, spreading the legs for balance, and using the arms and legs for propulsion.

25. BOARDING LIFESAVING CRAFT. a. Lifeboat or life raft.

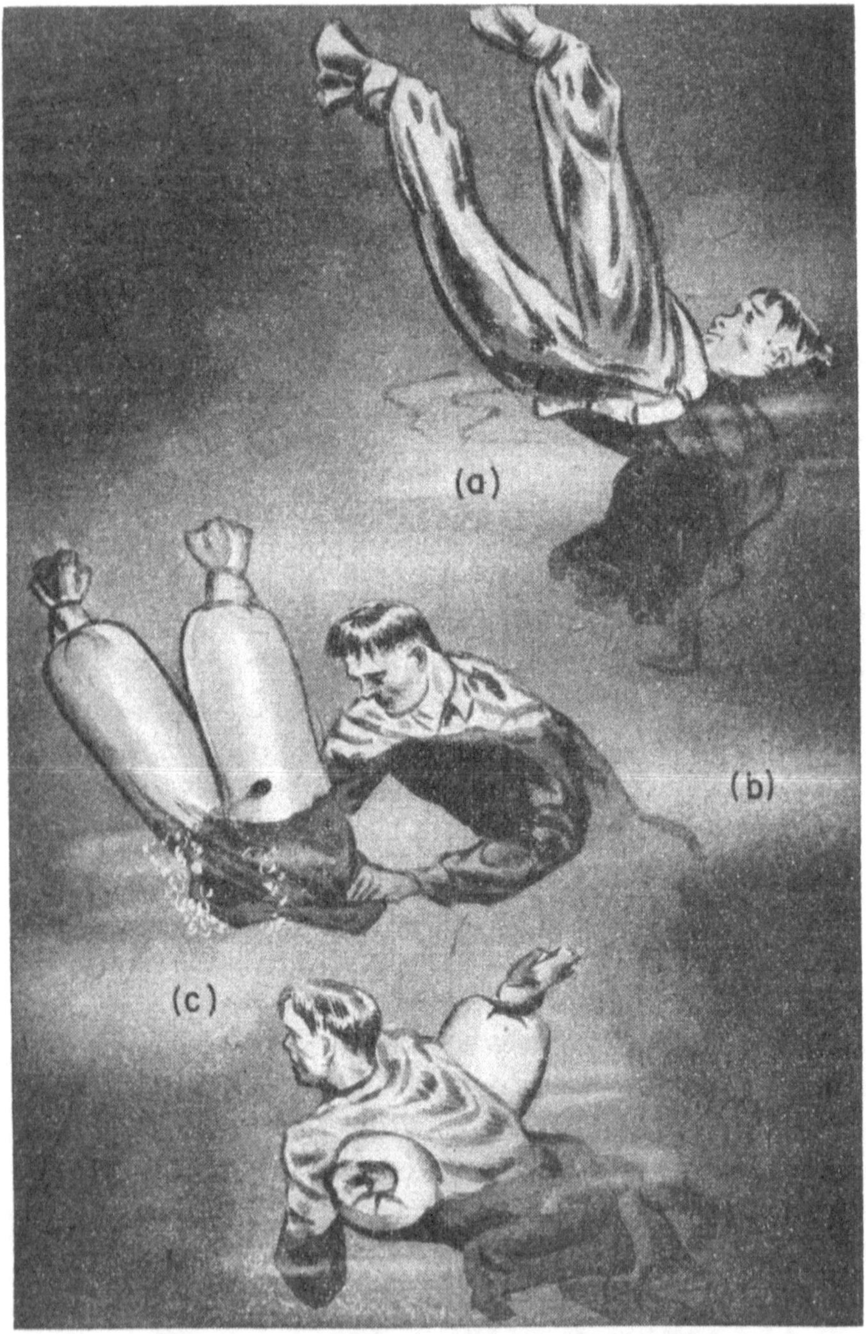

Figure 20. Using trousers for flotation.

(1) Because of their high sides and general shape, lifeboats should be boarded from the center of one side. Face the boat squarely, hook your arms over the side, wait for the next swell to raise your body, and, with a kick, roll into the boat.

(2) A man in a lifeboat or raft should aid a man in the water by lifting him above the edge of the craft until his body can bend at the waist. This brings his head and shoulders into the craft. The rescuer then grasps a leg and pivots the rest of the body into the boat.

b. Rubber boats. A lone survivor should board a rubber boat over the bow or stern. If there are two or more men, entrance over the side is recommended. (See fig. 21.) One man clings to the side of the boat. The other, on the opposite side, places one arm in the boat and locks it against the side. He then grasps the top of the side with his other hand, lifts his leg on the same side as the arm in the boat, and hooks his foot inside the boat. As the next swell lifts the boat, he pulls with the arm and leg in the boat, kicks down with the foot in the water, and rolls into the boat. The other survivor then boards the same way, and the man in the boat provides leverage by staying on the opposite side.

26. SWIMMING. Military swimming means swimming in full clothing and equipment to reach an objective in fit condition for action. If shipwrecked far from shore, the soldier does not try to swim to shore but tries to keep afloat until he is rescued. Basic water skills which conserve energy and which best in-

Figure 21. Method of boarding rubber boat.

sure buoyancy for long periods are discussed briefly in the following paragraphs.

a. Submerging and floating.

(1) *Submerging.* Initially, a man must be taught to submerge in shallow water and keep his eyes open. He must learn by experience that it is difficult to stay under

water because the buoyance of air trapped in his chest forces him to the surface. By expelling part of the air he can submerge without effort. He learns how to expel air through his nose under water, and to gulp air through his mouth above water. In deeper water he is taught how to bob up and down. Bobbing is performed by raising the arms abruptly, sideward and upward, to submerge; and lowering the arms abruptly, sideward and downward, to rise.

(2) *Floating.* Floating is the best way to conserve energy. Anyone can float, either motionless or with a slight movement of the arms or legs. To increase body buoyance the chest is expanded as much as possible. After taking a deep breath, further expansion can be accomplished merely by contracting (tightening) the abdominal muscles. Floating, combined with particular arm and leg movements to give propulsion, is swimming.

(*a*) *Back float.* This, the best relaxing float, can be accomplished with legs together or apart and arms extended to the side or overhead. The chin and chest should be raised as high as possible. With legs together and arms by the side there is a tendency for the legs to sink and drag the body under.

(*b*) *Jellyfish float.* The body is doubled up by bending the legs, pulling the knees to the chest, and dropping the head on the chest. Arms clasp knees. In this position the body will roll forward until only the back is visible above water. The jellyfish float is used mainly when removing clothing.

(*c*) *Prone float.* This is the basic position for all prone swimming strokes. It is performed by lying face down on the water, arms and legs extended. It is used with the face submerged, eyes open to see under the water or with the head high, to observe above.

b. Prone strokes.

(1) *Dog paddle.* This is done from the prone float position by using the arms and legs in exactly the same manner as when climbing a ladder. The hands, slightly cupped, fingers joined, thumb along the forefinger, reach forward and pull on the water. Then reach as far forward as possible without exertion and pull back on the water. The legs push back on the water by being doubled up and extended to the rear. Left leg and left arm, right leg and right arm, work together. Since legs and arms do not break surface, this is a useful stroke for silent swimming.

(2) *Breast stroke.* The breast stroke, like the dog paddle, is an extremely easy and relaxed stroke. It can be used in all types of water, calm or choppy, and should be done with the head held high for better observation. It is also used for underwater swimming. Start with the prone float and pull arms to sides horizontally; when arms reach shoulder level, start to draw in legs. Arms lunge forward; legs lash out to side straddle hop and then

return to starting position. By splashing with hands and arms on the forward movement, the breast stroke may be used to push away debris and thin surface oil and flame. It is a good stroke to use when carrying equipment on the back.

c. Treading water. Treading water consists of standing upright in the water with the feet going through the motions of climbing a ladder. It should be sufficient to keep the head high out of the water. If desired, the hands may be employed to push down on the water. Treading water is extremely useful when stopping to observe.

d. Side stroke. The side stroke may be performed on either side and hence is especially useful if either arm is disabled. Lie on side, one hand lower than other; execute deep knee bend with legs and knees together. Thrust legs backward and apart; then close them fully extended. At the same time thrust lower arm forward and upward and upper hand downwards and backward across body. Pull lower arm back in under chest and carry upper arm forward. Legs return to deep knee bend. The side stroke may be used to carry equipment above water or to tow an object or another man. If an item of equipment is too heavy for one man to carry, two men, using the side stroke, can carry the object with their free arms.

e. Elementary back stroke. The elementary back stroke is a relaxed stroke derived from the back float. The legs do the inverted frog kick or scissors kick. Arms are raised shoulder high, elbows straight, and then carried to the side, pushing the water towards the feet.

f. Jumping and shallow diving. Refer to paragraphs 16*b* and 21 for information on jumping and paragraph 22*a* for information on shallow diving.

g. Underwater swimming. In underwater swimming, either the dog paddle or breast stroke is used, the latter being preferable. The head is held high to see ahead or to see the surface of the water. Underwater swimming is used to escape oil, surface flame, or debris (par. 21*b*(2)).

h. Swimming to shore.

(1) When it is necessary to swim ashore from a shipwrecked vessel, every soldier should know how to protect himself. As he approaches shore, he should take up the breast stroke or dog paddle and survey the conditions of the beach (rock, sand), power of the surf, undertow, tide rip, and existing currents. Surf is the breaking of the waves on a beach. Undertow is the strong current beneath the surface that sets seaward after the wave has passed over; it is only within the first line of breakers. Tide rip is a condition set up when an outgoing tide opposes and slips under an incoming tide. The line of opposition is usually identified by a line of frothing water.

(2) After surveying the beach, the swimmer should swim opposite the point selected for landing or, if a current exists, he should swim to such a point that by going diagonally across the current the landing spot will be reached. He should remove the life preserver and all other means of support and discard them, if possible. A weak swimmer may hang on to them but he should always be ready

to let go, as retention of the life jacket and gear will prevent ducking the waves on the approach to the beach and will also make the swimmer more liable to be thrown on the rocks or violently onto the beach because of buoyancy created by the life preservers. The swimmer should use the breast stroke, side stroke, or dog paddle, looking behind to see the waves and looking ahead to check the direction. He should bob under the breaking waves, then come up and swim toward shore, continuously looking behind for the next approaching wave. He should not fight the undertow and should remember that it exists only for a short distance out below the surface. A strong push on the bottom with the feet will push the swimmer to the surface immediately after being rolled by the undertow, putting him in position to collect himself again and try for another landing. He must avoid panic and conserve energy. The wave will advance the swimmer more than the undertow will carry him back. He should swim easily until the next breaker comes along and then come in on the forward swell following the breaker. If he can stand on the bottom easily, he should watch the waves more carefully. They vary in size. He should go under the large ones and walk toward shore with the small ones, bracing his feet against returning undertow. He must always be patient even though making only a few feet at a time.

(3) In landing on a rocky shore, the swimmer

must be careful to avoid being thrown upon the rocks by a breaking wave. The best possible way to avoid wave action is to swim under water as deeply as possible where the water is calmer.

(4) The weak swimmer must remain outside a breaking surf with his life preserver until help arrives. If no aid is available, he must swim along the shore outside the surf and with the current, looking for an inlet to a river or bay, a long jetty, or a point where the surf breaks only when close to shore.

CHAPTER 4

OPERATING LIFESAVING CRAFT

Section I. COMMAND ABOARD LIFESAVING CRAFT

27. DETERMINING COMMAND. Command aboard a lifeboat is assigned by the master of the vessel to a member of the ship's crew and takes precedence over rank. If the boat commander becomes a casualty or if assigned commander is not present, the next senior officer or senior noncommissioned officer of the vessel's crew aboard the lifeboat assumes command. In the absence of these, the person most experienced in seamanship and the handling of people should take charge.

28. RESPONSIBILITIES OF COMMAND. The responsibilities of a boat commander are great. He should appoint at least two others as his assistants. Almost everything depends on the bearing and conduct of those in charge. They must be able to assume responsibility, enforce strict discipline, assign jobs, deal with emergencies, and take charge of rations, navigation, and boat work. Other specific command duties are to—

a. Set a definite course and maintain it.

b. Take charge of first-aid equipment and supervise administration of first aid.

c. Ration water and food; schedule the number of meals and time for each.

d. Assign tasks to all men except those severely exhausted or seriously wounded.

e. Place a man in charge of all water and provisions as a precaution against contamination, spoiling, and pilferage.

f. Place a man in charge of all arms and ammunition as a precaution against mutiny or conduct of a person become insane.

g. Arrange living and sleeping accommodations.

h. Examine all equipment aboard for serviceability; and supervise repairs. Divide all equipment, whether general or personal, to obtain an equal share of comfort.

i. Arrange suitable diet for each person per day —depending on the provisions aboard.

j. Maintain morale and faith. Schedule and conduct or supervise regular periods of worship if circumstances permit.

Section II.
ROWING AND SAILING LIFESAVING CRAFT

29. ROWING TECHNIQUE. Rowing a lifeboat may be necessary to get away from a sinking ship, to pick up survivors, or to make headway. It is difficult to row in a heavy sea. Men can row for about 15 minutes and, by making every stroke count, can move perhaps half a mile. Discipline and coordination are essential.

a. Hints.

(1) Be sure you do not pull on the oar before it is in the water.

(2) Keep your eye on the stroke oar on each side of the boat.

(3) Put the weight of your body on the oar. In a boat properly rowed, a thrumming noise

issues from the oars and gives a sense of timing.

(4) Oars should be stored after use to prevent warping. Twisted oars are practically worthless for rowing.

b. Learning to row. Steps taken in rowing a boat are as follows:

(1) Sit facing aft, body square with the thwart.

(2) Make sure your feet are comfortable and in proper position on the stretcher. Adjust the stringer if necessary.

(3) Grasp the oar easily, with the palms down.

(4) With the blade vertical, wrists straight, body leaning forward considerably, begin the stroke.

(5) As the stroke begins, raise the handle and dip the blade. Let the body do the pulling on the first two-thirds of the stroke. The last third of the stroke should be completed by the arms. When the stroke is finished, the body should be in a nearly upright position.

(6) When completing the stroke, turn the blade to a horizontal position on the recovery. This is called feathering the oar. Drop the wrist as the blade leaves the water to present the upper edge forward and reduce wind resistance.

30. GENERAL SAILING PRECAUTIONS. Sailing a lifeboat should be left to experienced hands aboard. In their absence, every caution must be observed, since a lifeboat may turn sideways and capsize in a matter of seconds if it is not handled properly. Even when sailing under the direction of an expert, keep your eyes open for an accidental jibe—the sudden swinging of the boom from one side of the boat to the other. Jibes have injured men seriously and have thrown others overboard. In heavy seas, high winds, or whenever there is any doubt about the weather, no attempt should be made to rig sails.

31. USE OF THE SEA ANCHOR. In heavy seas or strong headwinds, the safest course may lie directly into the wind or seas. To keep the bow of the lifeboat headed into the wind or sea, the sea anchor may be used. The sea anchor consists of a conical canvas bag and an inner perforated container of oil (fig. 22). When this equipment is put out over the bow with the large opening foremost, it fills with water, creates a drag, and assists in holding the bow into the weather. The sea anchor should have a tripping line attached to the pointed end, so that it may be pulled aboard easily when it is no longer needed. Even though this equipment is used, a careful watch must be maintained for combers that may break at the wrong moment and swamp the lifeboat. If these heavy seas are anticipated in time, the lifeboat should be maneuvered to meet them.

32. HANDLING RAFTS, FLOATS, AND RUBBER BOATS.

a. General. Rafts, floats, and rubber boats are designed primarily to provide flotation for men in the water before they are rescued. It is too exhausting, even under normal conditions of wind and sea, to attempt any headway in a raft or float. Save your strength to keep afloat; do not attempt to make distance. Some progress can be made in rubber boats, but it is not advisable to attempt headway in these

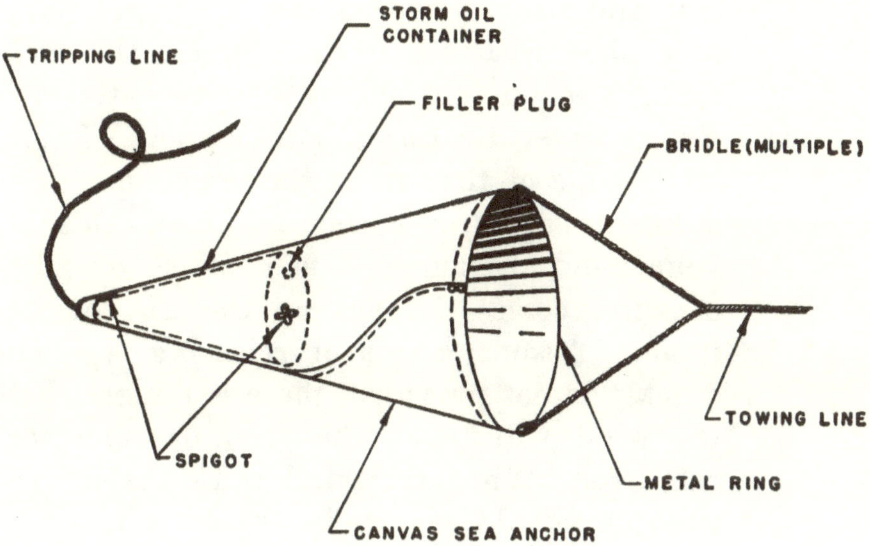

Figure 22. Sea anchor.

craft if land or an air or sea lane is distant. Keeping on the move rather than just drifting may help from a morale standpoint.

b. Effect of winds and current on rubber boats. With a little care, anyone can sail a rubber boat with the wind. A raft for carrying several men can be sailed successfully 10° off from the direction of the wind.

(1) *Preparation.* When the wind is blowing directly toward your destination, inflate the raft fully, sit high, rig a sail, and use an oar as a rudder. In a multiplace raft, rig a square sail in the bow, using an oar (with extension) as a mast and another as a cross bar. If the regular sail is not available, the waterproof tarpaulin or two thicknesses of parachute cloth will work. If the boat has no regular mast socket and step, put up the mast by lashing it securely to the front cross

seat and prop it with stays. The heel part of a shoe with toe wedged under the seat makes a good improvised mast socket.

(2) *Sail.* Never tie down both corners of the lower edge of the sail at the same time. Tie a line to the free corner and hold this in your hand. When a sudden storm or gust of wind comes, let out on the line or let it go. If some means of releasing the sail quickly is not provided, there is danger that the wind will break the mast or turn the raft over. When the wind is favorable, occupants should sit up in the boat to offer resistance to the wind.

(3) *Sea anchor.* When ocean currents are moving toward your destination but the wind is unfavorable, put out a sea anchor. Deflate the raft slightly so that it rides low in the water. Huddle low in the raft to offer as little resistance to the wind as possible. In the open ocean, a current can move the raft 6 or 8 miles a day.

CHAPTER 5

NAVIGATION FROM LIFESAVING CRAFT

Section I. DETERMINING DIRECTION

33. USE OF THE COMPASS. Most lifeboats and some other lifesaving craft will be equipped with a standard compass. If this compass is not in a fixed mount, it should be secured in a position where it can be read easily by the person who is steering the lifeboat. (See fig. 23.) In securing the compass, care should be taken that the fore and aft lubber line on the compass (fig. 24) is parallel or in line with the centerline of the boat. The compass in the lifeboat provides a means for steering a certain course, if possible, or for indicating the direction of your drift.

34. DIRECTION FROM THE SUN. a. By rising and setting of sun. Observe the times of rising and setting of the sun on the same day, or the time it sets one day and the time it rises the following morning. Divide by two the time elapsed between rising and setting. In the Northern Hemisphere, the answer, added to the time of the sun's rise, will give the hour when the sun is true south.

Example:

Sunrise . . . 0600
Sunset. . . . 1900

$$\frac{1900-0600}{2}=\frac{1300}{2}=6\tfrac{1}{2}\text{ hours}$$

Time when sun is true south . . . . 0600 + 0630 = 1230

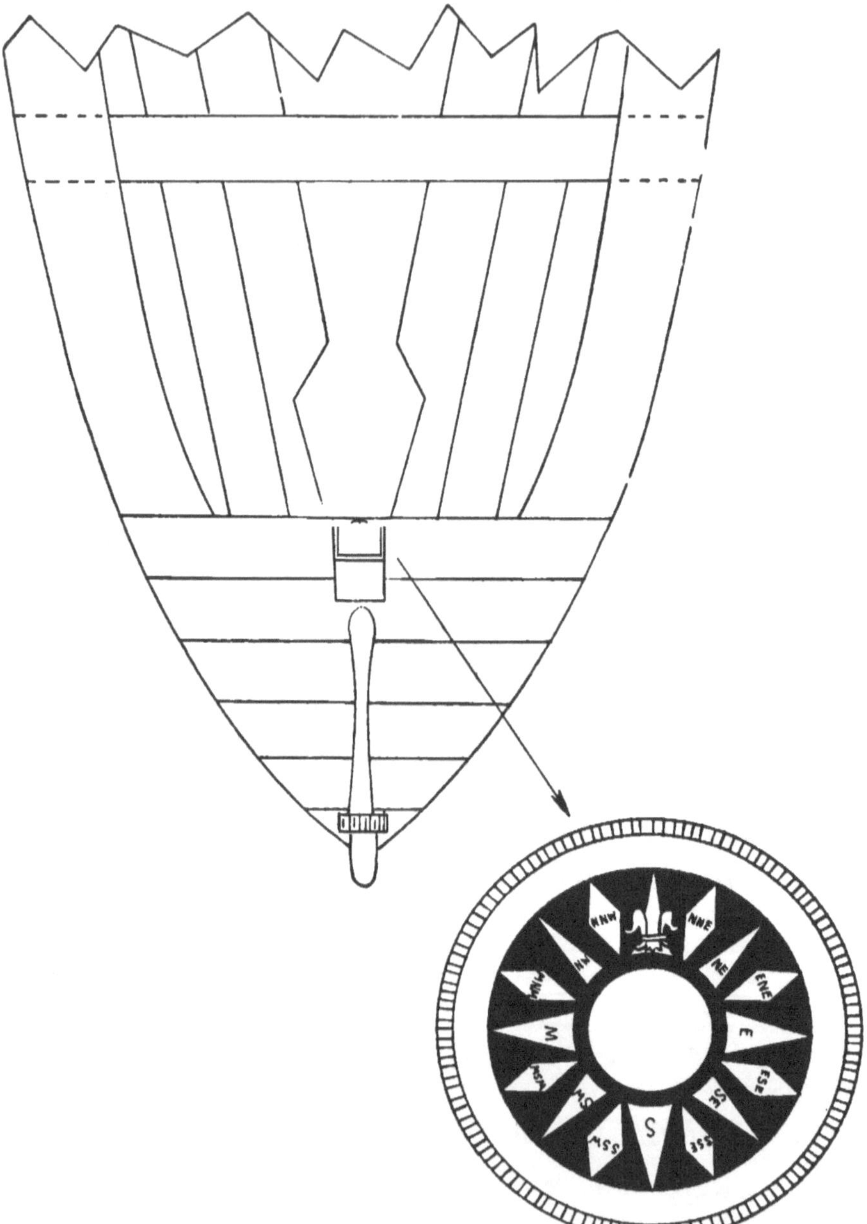

Figure 23. Position of mariner's compass in lifeboat.

The same procedure applies for the Southern Hemisphere, except that the direction to the sun is true north.

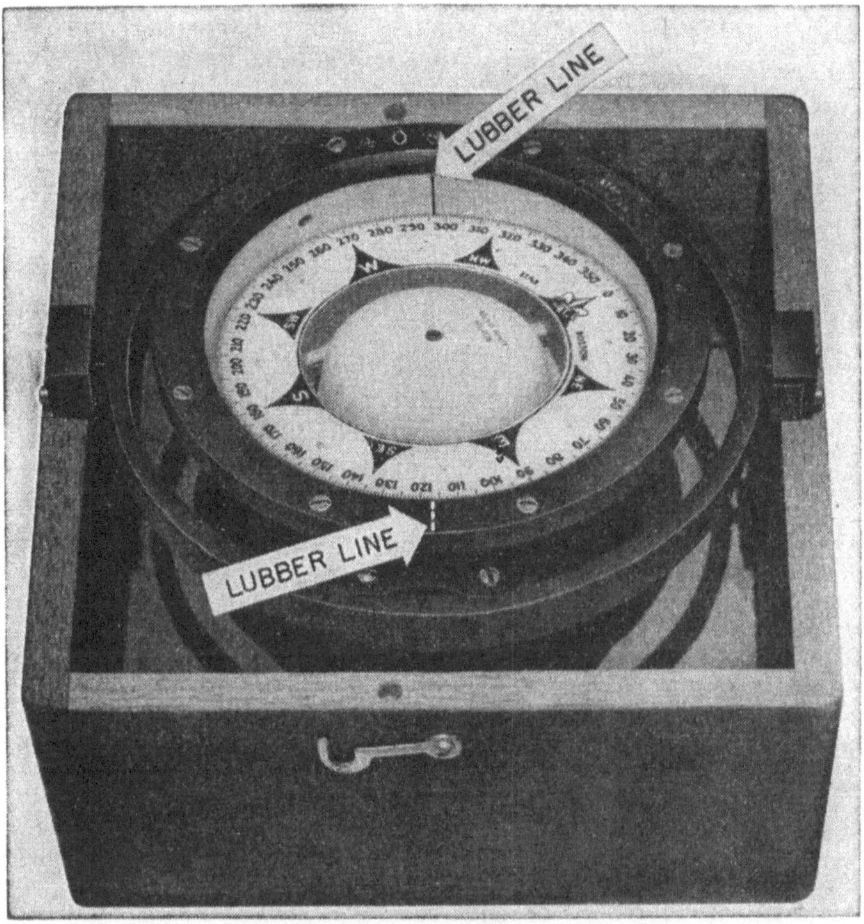

Figure 24. Standard compass.

b. By watch and sun. When the sun is visible, a watch can be used to determine true south or north with an error of less than 8°. This method is difficult when the sun is high; it is of little or no use in the Tropics. Furthermore, the watch must be on standard time. If it is subject to corrections for wartime, daylight saving time, or zone time, it must be set back accordingly.

(1) *Northern Hemisphere.* In the Northern Hemisphere, turn the watch face up, and

point the hour hand at the sun. To aid in correct pointing, hold vertically a pencil or other straight, slender stick so that it casts a shadow across the face of the watch. Rotate the watch to bring the hour hand into this shadow. Draw a line from the center of the watch dial through the midpoint of the smaller arc between the hour hand and 12 o'clock on the watch face (Ⓐ in fig. 25). This line points toward true south.

(2) *Southern Hemisphere.* In the Southern Hemisphere, point the 12 o'clock mark on the watch toward the sun. North lies halfway between this mark and the hour hand (Ⓑ in fig. 25).

35. DIRECTION BY THE STARS. a. Northern Hemisphere. In the Northern Hemisphere, the North Star (Polaris) is the best star for finding direction. This star is almost vertically above the North Pole and any sight on it is within 1° of true north. The North Star may be identified in the following ways:

(1) *By means of the Big Dipper.* The Big Dipper is a star group easily recognized by its shape (fig. 26). The two stars forming the side opposite the handle are called pointers, because a line through them always points to the North Star. From the lip of the Big Dipper to the North Star is about five times the distance between the pointers. Anyone having difficulty in locating the North Star can do so by using the fingers (fig. 27). Hold one finger in front of the eye and adjust its distance from the eye until

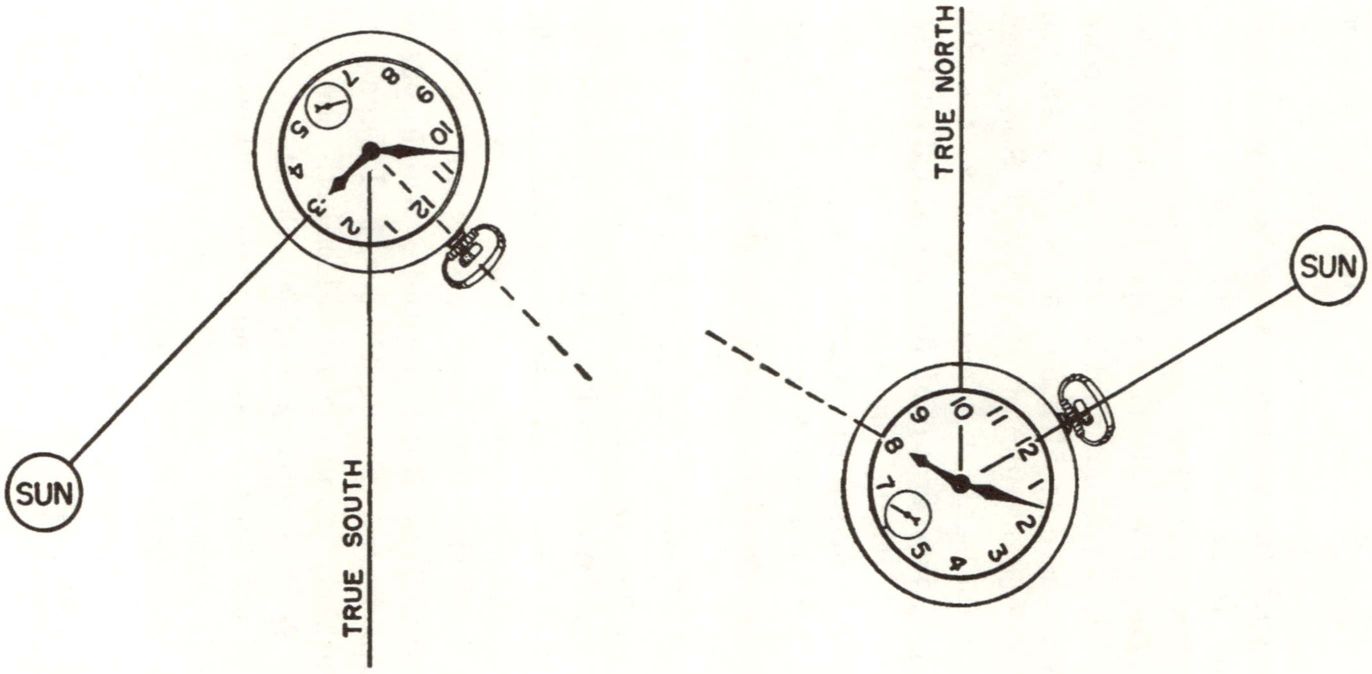

Ⓐ NORTHERN HEMISPHERE Ⓑ SOUTHERN HEMISPHERE

Figure 25. Use of watch and sun to determine direction.

one pointer is at each side of the finger. Add five more fingers. The North Star then is just outside the last added finger and on a line with the pointers. Once identified it is easily recognized by its brightness in comparison with other stars nearby.

(2) *By means of W (Cassiopeia).* When the Big Dipper is not visible, another star group may be used to identify the North Star. On the opposite side of the North Star, and about the same distance from it as the Big Dipper, is a group of five stars, called Cassiopeia, forming the letter W (or M if the group is above the North Star). The relation between the North Star and W (Cassiopeia), shown in figure 28, should be memorized.

b. Behavior of stars. At the North Pole, the North Star appears directly overhead, and both the Big Dipper and W are visible and seem to rotate around the North Star. As one goes south from the North Pole toward the Equator, these stars appear to lose elevation; they are seen nearer the horizon. The 40th parallel of north latitude, which passes through Pennsylvania, Spain, Greece, and Japan, is the most southerly point from which both the Big Dipper and W are always visible. South of this parallel, there are times when only one of these star groups is visible; therefore, it is necessary to be able to locate the North Star by reference to the Big Dipper or to W. (See fig. 26.)

c. Southern Hemisphere. In the Southern Hemisphere, the Southern Cross, or True Cross, is the most distinctive constellation. It has four stars. Those on

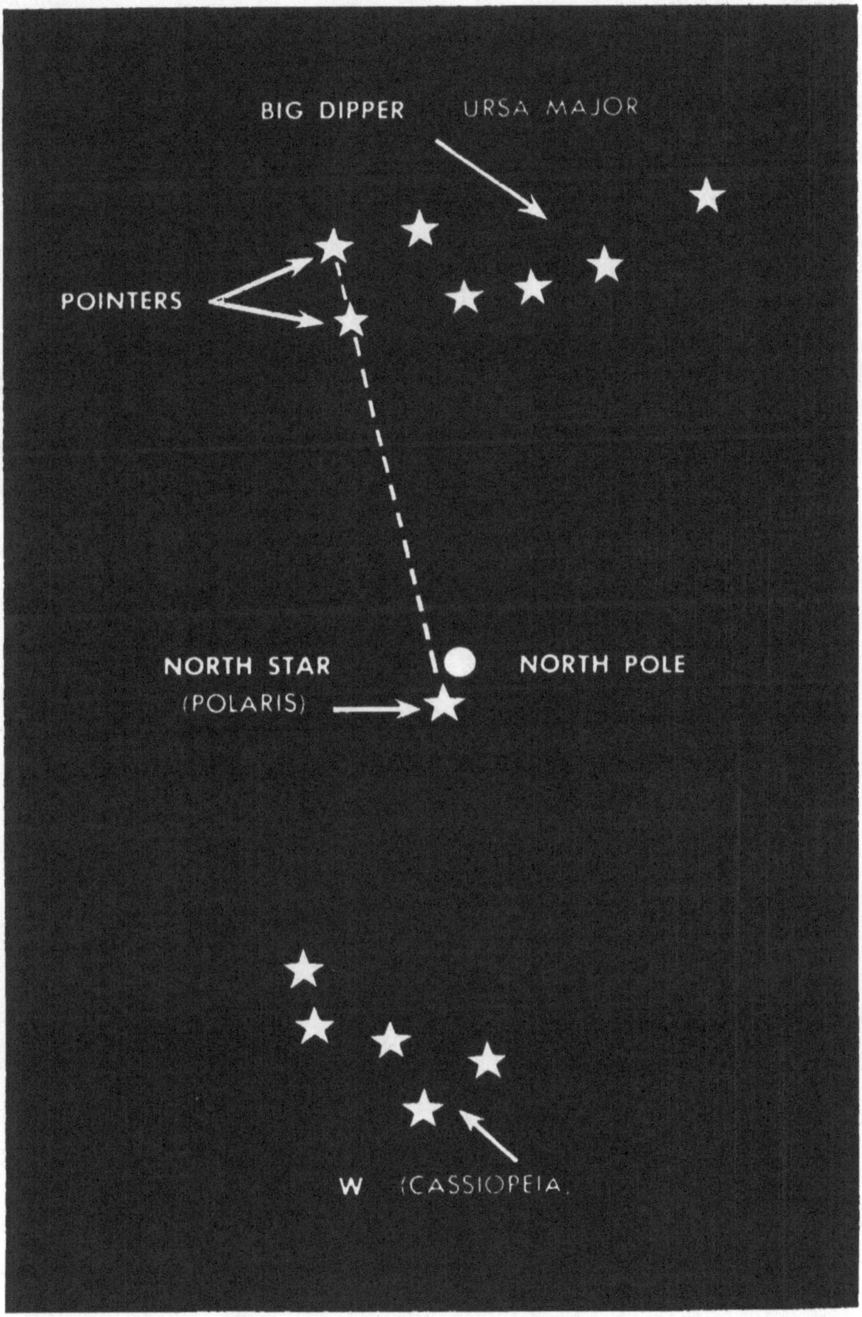

Figure 26. Relation of the Big Dipper and W (Cassiopeia) to North Star.

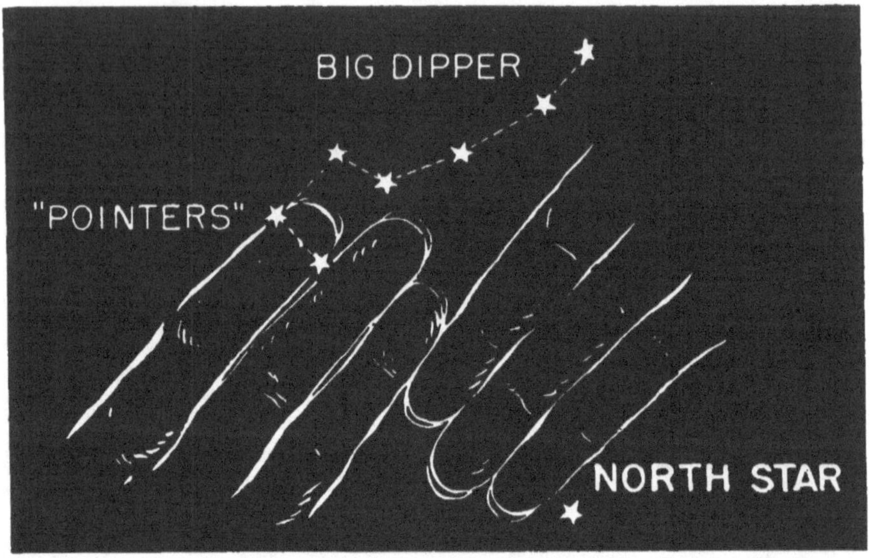

Figure 27. Using the Big Dipper and fingers to locate the North Star.

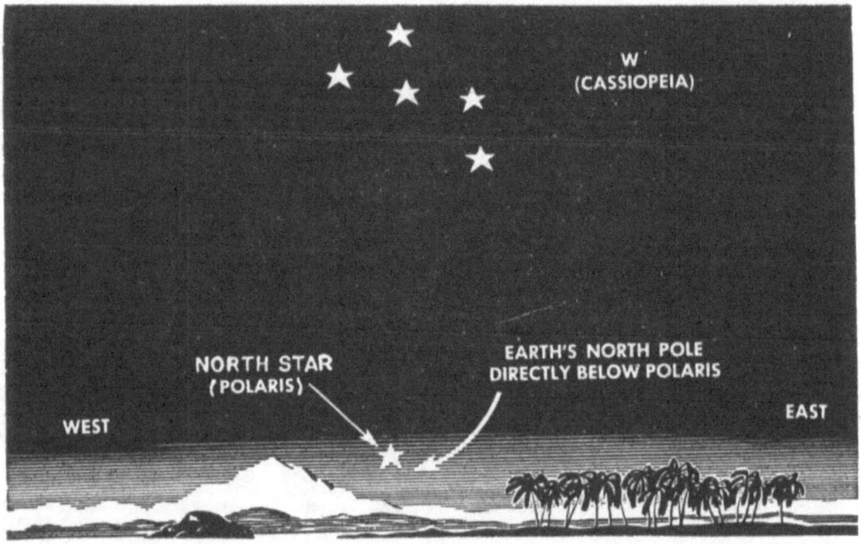

Figure 28. View of North Star and W from Equator.

the southern and eastern arms are the brightest in the heavens; those on the northern and western arms are bright but smaller. The True Cross should not be confused with the False Cross which lies to the west and has five stars which are more widely spaced and less bright. There is a region immediately above the South Pole which is so devoid of stars that it is called the Coal Sack. To the east of the True Cross are two bright stars which can be used along with those of the True Cross as guides to the location of the South Pole. Join the two stars with an imaginary line; then bisect this line with one at right angles. Draw an imaginary line through the axis of the True Cross. The intersection of this line and the one from the two stars is approximately the point above the South Pole. (See fig. 29.) From the vicinity of the Equator, both the Southern Cross and the Big Dipper may be visible, sometimes only one of them. When both are visible, they are about equally high above the horizon, but in opposite parts of the sky. As one moves south from the Equator, the Southern Cross becomes visible for a greater part of the night. South of the 33d parallel of south latitude, which runs through Uruguay, Cape of Good Hope, and southern Australia, the Cross is visible all night. For night observation south of the Equator, where the Southern Cross is not continuously visible, the following procedure may be used: Determine direction just before the Southern Cross disappears, by methods already described; select a star in the vicinity of the South Pole, preferably one just rising, and memorize the appearance and position of this star; use it to maintain direction for the remainder of the night.

Section II. PHYSICAL INDICATIONS OF LAND

36. GENERAL. Certain signs may aid the helmsman in steering toward land. Of themselves, these signs

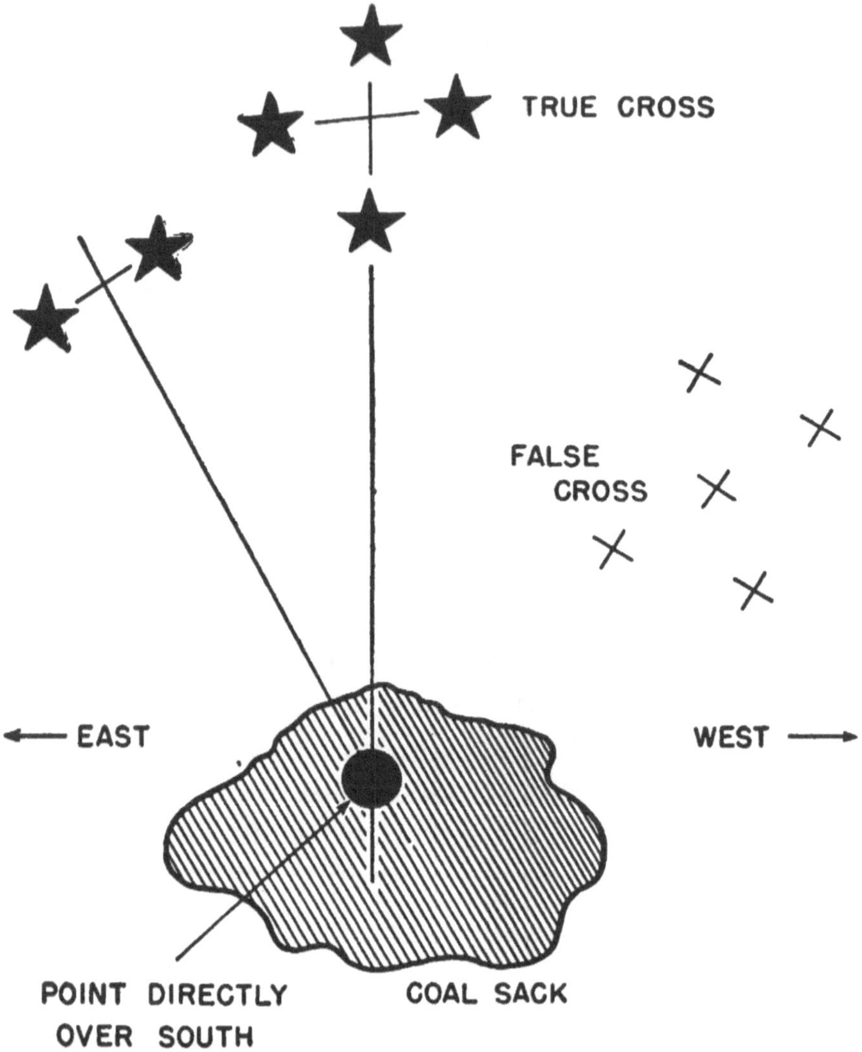

Figure 29. Relation of Southern Cross and False Cross.

are not positive evidence of the proximity of land; correlated with other observations, they strengthen the probability of finding land nearby.

37. INDICATIONS OF CLOUDS. Clouds and certain distinctive reflections in the sky are the most reliable indications of land.

a. Atolls. Small clouds hang over atolls a little to the lee side. Color of a lagoon is sometimes reflected from the clouds and indicates an atoll beyond .the horizon.

b. Shoals. Small clouds may also hover over coral patches and hidden reefs, thus providing a warning of shoals.

c. Fixed clouds. Fixed clouds or cloud crests often appear around the summits of hilly islands or coastal land. They are easily recognized, as moving clouds pass by them.

d. City lights. Lights from cities are usually reflected in the sky, especially by high clouds.

e. Lightning. In tropical regions, lightning from one particular direction in the early hours of morning usually indicates mountainous territory.

f. Arctic regions. In polar regions, a sharply defined patch of brightness in otherwise gray sky is a sign of areas of floe or shore ice in the midst of open water.

38. INDICATIONS BY SOUND. Sound from land is affected by the strength and direction of the wind. To be useful to the navigator, this fact must be applied to any sound from land. By shutting the eyes and turning the head to get equal volume of sound in each ear, it is possible to obtain close approximation of the bearing of the sound.

a. Birds. Continued cries of sea birds from one particular direction signify roosting place on land.

b. Fog. In fog, if a ship's whistle or siren is heard, the vessel is moving; but if a bell is heard at regular intervals, the sound is coming from a ship at anchor or from a bell buoy.

39. OTHER INDICATIONS OF LAND.

a. Birds and insects. An increase in the number of birds and insects indicates land nearby.

b. Odors. Land odors are carried seaward by the wind. Detection of such odors in fog, mist, rain, or at night is very important, as it is possible to drift past a nearby shore without seeing it.

c. Seaweed. Seaweed is usually found in shallow water and its presence denotes nearby land. However, make allowance for strong currents that may have carried it from the original source. An exception is the Sargasso Sea, a region of the North Atlantic Ocean (lat. 16°–38° N., long. 30°–50° W.) where some of the surface is covered with floating gulfweed.

d. Ice. Bay ice, usually smooth, flat with jagged edges, and whiter in appearance than pack ice, indicates a frozen inlet nearby, especially if the pieces are close together. If the ice is scattered widely with its edges crushed, land may be quite distant. However, the current or wind will indicate the direction of land.

CHAPTER 6

ADRIFT AT SEA

Section I.
SUSTENANCE ABOARD LIFESAVING CRAFT

40. CARE OF EQUIPMENT. **a. Boat equipment.** Lash down everything aboard. Do not discard anything unless you are sure it will be of no further use. Keep all equipment as dry as conditions permit. Every effort should be made to dry the boat and keep it dry.

b. Clothing. As soon as possible, squeeze out all wet clothing. Do not disrobe entirely unless the weather is warm and dry, with a moderate wind. Dry your clothes layer by layer.

41. WATER. **a. General.** Water is the most important item for survival. Man can exist only about 7 days without it. Loss of body moisture is hastened by heat and exercise. *Do not drink salt water,* as this will cause diarrhea, weakness, and unbearable thirst. Delirious men may have to be restrained from drinking salt water. *Never drink urine.* If water is not available, do not eat. Elimination of food wastes absorbs water from the kidneys and decreases water in the body.

b. Sources.

(1) *Boat supply.* This is the primary source of drinking water. To keep water from freezing in cold climates, remove the plug. Insert a stick in the container. The lower end of the stick should be weighted and upper end

should protrude a foot or more. Movement of the boat will keep the stick in motion so that ice will not form.

(2) *Canteens.* All canteens will be commandered by the commander of the boat and the water in them made a part of the boat's supply.

(3) *Rain water.* Use buckets, cups, tin cans, sea anchor, boat cover, sails, strips of clean clothing, and all canvas gear in the boat to collect rain water. Wash salt off equipment with the first fall of rain, as it is necessary for drinking water to be free of salt.

(4) *Ice.* In the cold regions icebergs are a source of fresh water. Sea ice loses its salt after a year and becomes a good source of fresh water. This "old" ice has rounded corners because of rains and thaws, is bluish in color, has a glare, and is splintered easily with a knife. In freezing weather, fresh water can be obtained from sea water. Collect some sea water in a container. The fresh water will freeze first, and the salt will collect in high concentration as slush in the core of the frozen piece. Remove the ice and throw away the slush. The melting ice will produce water sufficiently free of salt to sustain life.

(5) *Chemical kits.* Chemical kits to remove the salt and alkaline from salt water may be provided in lifeboats or rafts. Directions accompany the kits.

(6) *Other measures.*

(*a*) *Dew.* In certain areas, dew may form. It can be collected by rigging a tarpaulin with the edges turned up, allowing the water to accumulate in the middle.

(*b*) *Fish.* The chewing of small pieces of fish will provide some moisture and help to arrest thirst.

(*c*) *Condensation.* If all other means of obtaining drinking water have been exhausted, any metal container and lighted lantern may be used to obtain water by condensation. Remove one end of the container and submerge the closed end in a foot or more of salt water. Place the lighted lantern inside the container, on the bottom. Cover the open top, allowing only enough air to enter to keep the lantern burning. The heat will cause moisture to form on the inside of the container. This can be soaked up with a rag and squeezed into a cup.

c. Storage. Water should be stored in every container that can be used for this purpose. It may be stored in carbon dioxide life jackets and in the air tanks of the lifeboat.

d. Issue. An inventory will be taken of all water aboard. Plan the issue of water with regard to all circumstances present: total supply of water, number and condition of the survivors, time likely to be adrift, chances of replenishing supply by rain, and output of chemical kit. The average daily ration is 18 ounces—three cups. Control of issue starts immediately and continues until survivors are rescued.

e. Hints.

(1) After a rain, drink your fill slowly over an hour or more. Kidneys will not then waste

water. Always drink water slowly and in small amounts.

(2) Do not drink the liquid in the compass. It is poisonous.

(3) Chewing gum or sucking on a button helps reduce thirst.

(4) When water is scarce, just moisten lips and throat. Hold water in your mouth and gargle before swallowing.

(5) Before abandoning ship, try to get a drink of water. If you drink a lot before abandoning ship, you should be able to go 24 hours without requiring another drink.

(6) Diabetics and those suffering with fever require an extra ration of water.

(7) Distinguish between true and artificial thirst. True thirst is attended by a burning irritation and complete dryness in the mouth and throat. Artificial thirst is created by the thought of water or its need, or by eating food or drinking water containing salt or sugar.

(8) Avoid drinking alcoholic beverages.

42. FOOD. a. General. A responsible person must be put in charge of all food supplies. He must divide all food fairly and schedule meals. Before any food is distributed, a complete inventory of provisions should be taken. Food should be checked periodically, especially on dry days, to see if anything has spoiled.

b. Sources.

(1) *Lifesaving craft supply.* This is the principal source of food supply.

(2) *Emergency kits.* Your personal emergency kit is an excellent place to store nourishing, concentrated food for use during an emergency. Such food may be jam, which contains 1,300 calories per pound; walnuts, 3,300 calories; dried coconut, 2,600; and chocolate candy, 2,500.

(3) *Birds.* All birds are good to eat, cooked or raw, and their blood and livers are also edible. Entrails, head, and feet make good bait. The feathers may be used to make fishing jigs, or they may be stuffed under your clothing for warmth. Sea birds are usually difficult to catch. Normally, they follow schools of fish, and you can attract them to your craft by tossing a piece of fish in the air. You can catch a bird by grabbing at it, but slingshots, dip nets, or harpoons are more effective. Birds may also be ensnared on a baited hook dragged through the water or thrown into the air.

(4) *Fish.*

(*a*) *Edibility.* Practically all freshly caught sea fish are palatable and wholesome, cooked or raw. In warm regions fish should be bled and gutted immediately after catching. Fish not eaten immediately should be cut in thin, narrow strips and hung to dry. A well-dried fish will stay good for several days. Fish not cleaned and dried may spoil in half a day. Never eat a fish that has pale, slimy gills, sunken eyes, flabby skin and flesh, and unpleasant odor, or whose flesh remains indented when pressed by the thumb. Good fish should have pink or red gills, bright clear eyes, firm flesh, and be free from stringy slime. Sea fish should also have a salt water tang or clean, fishy odor. Eels are fish and good to eat, but do not confuse them with sea snakes. Unlike eels, sea snakes, found in the Pacific and Indian Oceans, have scales and swim on the surface of the water.

(*b*) *Poisonous fish.* Poisonous fish are found in the Tropics. The bodies of these fish are covered with rough or spiny scales, thorn-like spines, or bony plates. In one poisonous variety the skin is naked or is strewn with soft spines or bristles which look like hair. None has the ordinary scales found on bass, trout, snappers, groupers, and goldfish. Follow this rule: If it does not look like an ordinary fish, if it has unusual appendages, or an unusual looking mouth, lacks teeth, is not covered with ordinary fish scales, let it alone. Remember that fresh, nonpoisonous, salt water fish can be eaten raw; fresh water fish cannot.

(*c*) *Turtles.* The whole meat, blood, and juice of a turtle are edible, but the shell, stomach, and kidneys are not. Hot sun brings out of turtle fat a clear oil into which food may be dipped. Turtles may be snagged with a hook or grapple in the leg, neck, or edge of the shell. They can be harpooned through the shell or shot in the head. Even after a turtle's head has been cut off, the mouth may bite and the claws may scratch. The turtle should be consumed immediately, as turtle meat spoils rapidly.

(*d*) *Crabs and shrimp.* Crabs, shrimp, and small fish are often found in seaweed. To get these aquatic animals, pick up a bunch of seaweed and shake it out over the craft. Discard all jellyfish, as they are poisonous, and if your water supply is low, discard the crabs, as they cause thirst.

(*e*) *Fishing.* Improvised fishing rigs and supplies can be assembled.

1. Hooks may be made from items with points or pins, such as nail files, collar insignia and campaign ribbons, or from bird bones, pieces of wood, and fish spines.

2. Fishlines can be made from pieces of tarpaulin or canvas by raveling the threads and tying groups of three or more together in very short lengths. This will make a general fishline that will stand about a 20-pound pull. Parachute shroud lines, shoelaces, or thread from clothing may also be fashioned into lines.

3. Save the entrails, except the liver of fish, for bait. A pearl button is a good substitute for bait.

4. Besides the conventional methods of fishing, fish may be caught with the bare hands, after they have been enticed to the craft by dangling something in the water. You may attract fish at night by using lights or, in the sunlight or moonlight, by using a mirror or other shiny object. Flying fish can often be caught by hoisting sail or by hanging up a tarpaulin or piece of clothing. A light on the surface of the tarpaulin at night will help. The flying fish will hit the obstruction and fall into the craft. Other methods of catching fish include stunning them with an oar or stick while they are feeding, using a slingshot or improvised net, and, as a last resort, shooting them.

5. Wear gloves if you have them. Do not wind your line around your body or make it fast to the craft. Have another man hold the end of the line while you are fishing.

(5) *Seaweed.* All seaweed is edible. However, it is usually very salty and should not be eaten unless an abundance of water is available.

c. Issue. Control of issue starts immediately and continues until rescue.

(1) *Ration calculations.* To calculate rations, first estimate the number of days before rescue is expected. By dividing this number into the amount of each item of food, the daily ration of each is found. In a lifeboat loaded to capacity, there are 56 ounces of food, or about 8,000 calories, for each person. Provisions include the following:

56 biscuits—1 biscut weighs ½ ounce.

226 malted milk tablets—19 tablets weigh 1 ounce.

(Suck tablets slowly—do not chew them.)

4 cans of pemmican—¼ can weighs 1 ounce.

(Pemmican is concentrated meat; eat pemmican and biscuits together.)

Example: If rescue is expected within 10 days, the ration for 1 day will be:

5 biscuits.

22 malted milk tablets.

2/5 of a can of pemmican.

This equals about 5.6 ounces per day, giving a diet of about 800 calories, sufficient to sustain life.

(2) *Eating rations.* Eat slowly and chew thoroughly. Birds or fish should be fairly distributed to supplement the regular diet. Special food should be kept for use when morale is particularly low.

43. CONSERVING STRENGTH. a. Wasting energy. Energy used aboard a lifesaving craft is not likely to be replaced by the rations provided in the craft. Do not waste strength by useless exertion or by the development of a bad frame of mind. Follow these suggestions:

(1) Do not exhaust yourself by getting excited.

(2) Do not sing or shout.

(3) Take mild exercise such as a short turn at the oars to prevent body from kinking up.

(4) In hot weather, work on the boat should be done before the sun is up.

(5) It is never justifiable to attempt to make progress by continuous pulling at the oars. Periods of 15 minutes at the oars with 1

hour's rest will permit steady progress for long periods with minimum exhaustion.

(6) In a warm climate, keep your clothes constantly wet with sea water during the day. As the sea water evaporates, it will cool your body, thereby reducing perspiration and the evaporation of water in the body tissues.

b. Sleep. Sleep regularly. If you feel cold in a lifeboat, crowd together under a canvas cover forward or in a sailcloth or blanket. In calm weather, make more room in the boat by lashing oars and spare gear outboard along the gunwale. In a float, it may be dangerous to drop off to sleep because of the risk of drowning. In extremely cold weather, stay awake as long as possible.

44. PROTECTION AGAINST WEATHER. a. Cold winds, rain, and spray. If canvas hood and side spray curtains are available, put them up as soon as possible. In freezing weather a blanket dipped in water and allowed to freeze will provide shelter against spray and wind. Sometimes these measures will not give sufficient protection and you will have to share blankets and huddle together to keep warm. In wet weather, keep clothing on even if it is wet. At a favorable moment, dry it out as much as possible. Exercise your toes from time to time to increase circulation.

b. Sun and heat. Rig up an awning, if possible, and try to provide some cover for the man at the tiller. Do not take off too many clothes; they will protect your skin against sunburn. This also applies to legs and feet. Even in cloudy weather you can get badly sunburned. Protect eyes from glare of sun by improvising an eyeshade. Tie a cloth or bandage over nose; this will cut off glare from the water when you are looking straight ahead.

45. WATCH AND LOG. a. Keeping watch. Maintain a continuous watch. This is similar to sentry duty. The sentry looks for sudden changes of weather, rescue parties, signs of land, and leaks; steers the craft to prevent its capsizing; maintains the designated course or changes it as necessary; and informs the commander of the craft of all that has occurred during the watch. This duty is rotated and should be shortened as necessary when the craft's company becomes exhausted.

b. Maintaining a log book. Detailed information on events occurring during abandonment and while adrift, with accompanying comments and suggestions for improving methods and equipment, is entered in the log. A running daily account of the course, speed, weather, etc., aids in navigating and calculating the position of the craft. An inventory of provisions and water is entered, followed by a daily record of what is consumed. All information may be entered on the backs of available charts, on paper or in books, or even on the side of the boat itself.

46. MENTAL HEALTH. a. General. High morale will increase your chances of survival by keeping you in the best possible physical, as well as mental, condition. To maintain high morale, the mind and body must be occupied, if only with trivialities. Here are a few suggestions:

(1) *Discussions.* Lengthy discussions or debates will help pass the time. Swapping informa-

tion on each other's jobs or teaching subjects to the group is useful.

(2) *Recreation.* Do not overlook books, magazines, games, musical instruments, or a portable radio when abandoning ship. Tell all kinds of stories. Invent new things. Busy your mind with puzzles and riddles.

(3) *Tobacco.* Cigarettes are valuable at sea in steadying nerves and keeping minds balanced, especially during watch at night. The person in charge should confiscate all tobacco and then ration it. Equality is important, but the judicious use of special foods or an extra ration of water will raise morale.

(4) *Keep moving.* As long as the winds permit you to go in the general direction of your track, keep the craft moving. All will feel better if they know headway is being made.

(5) *Services.* Worship services should be encouraged. Proper burial of the dead, using a small flag and a brief prayer and service, is also important.

b. Activity. Use your wits and good sense to overcome anxiety, which may arise from worry, misconception, and lack of activity. By keeping account of the time and your appropriate whereabouts, you will lessen your anxiety considerably. If you have a watch, keep it wound and cared for. Devise a method of keeping track of the days. Keep busy.

c. Tricks of the mind. Isolation, fatigue, exposure, and extreme hunger and thirst can cause your mind to have hallucinations or hear imaginary noises. Clouds may sometimes look like ships or icebergs, though such mirages are usually temporary. The delusions may lead you to harm yourself, unless you are on the alert to recognize them for what they are.

d. Keeping faith. Whatever your belief, try to strengthen it while adrift. Fight for faith and hold on no matter what the circumstances may be. You must have faith in someone or something. Have faith in yourself and your ability to meet every situation no matter how difficult.

e. Will to live. The desire to live is an outstanding factor in this battle of survival which is a challenge to anyone. Keep your head and your mind occupied. When you find yourself in the water or adrift in a boat or on a raft, meet the challenge with a normal curiosity for what the next day holds in store. There will always be people trying to locate you.

Section II. FIRST AID AT SEA

47. GENERAL. This section covers only first aid generally necessary after disasters at sea. The discussion of cases and their treatment is intended for survivors of a shipwreck who must attempt to give medical care to fellow survivors in the cramped, exposed, quarters of a lifeboat, with the emergency first-aid equipment usually available. To use the following information to the best advantage, you must know the general principles of first aid and their application covered in FM 21–11. With all cases remember to use common sense. No two cases are alike and people react differently under hardships.

48. FIRST-AID KIT. **a.** Each lifeboat contains a first-aid kit. Usually, the following items are packed in a watertight, metal box:

7 4-in. compress bandages.
2 2-in. x 6 yd gauze bandages.
32 waterproof adhesive compresses.
1 40-in. triangular bandage.
100 aspirin, phenacetin, and caffeine tablets.
10 ammonia inhalants.
1 tourniquet.
1 forceps.
1 scissors.
12 safety pins.
3 vials tincture of iodine.
3 eye-dressing treatments.
4 tubes foille for burns.
135 benzedrine sulfate tablets.
100 phenobarbital tablets.

b. On each package of medicine or bandages, and on the inside cover of the first-aid kit, are instructions for the use of the contents. Read these instructions, follow directions, and conserve the contents of the kit.

49. APPLYING FIRST AID. a. General.

(1) Help the survivor from the water and place him in a reclining position with head low and feet raised. Examine him for injuries, swelling, immersion foot, burns, frostbite, numbness, paralysis, shock and any internal pain or tenderness. Handle him gently. Keep him warm but do not apply heat directly to his body. After making him as comfortable as possible, and if his condition permits, ask him questions as to the period of exposure, underwater explosions, general conditions during exposure, the amount of sea water

drunk, if any, and the amount of food and fresh water he has had.

(2) Lash weak or badly injured men to the boat to prevent their rolling about. If a flat surface is needed, several oars can be laid side by side in the boat or across the gunwales. Life preservers make a satisfactory bed. Post a man to prevent the sick from attempting to go over the side; they sometimes imagine they are back home or in the ship. Humor them at all times. In case of death, the clothing and equipment of the deceased should be removed before disposing of the body.

b. Treatment of specific cases.

(1) *Shock.*

(*a*) *Cause.* In every severe injury the body suffers from a certain amount of shock. This often is more serious than the wound itself and may cause death.

(*b*) *Symptoms.* Pallor, rapid and weak pulse, and nausea are the symptoms. Breathing is irregular and similar to sighing. The body may be cold and clammy and chills may be present. The eyes may be glassy and have a fixed stare.

(*c*) *Treatment.* Keep the patient in a horizontal position with feet elevated and head low except when there is an injury to the head. With a head injury the head must be elevated. Keep warm with sailcloth, blankets, or other means. Administer stimulants. Use ammonia inhalant from first-aid kit. Relieve pain. Give morphine, if available. Dosage: one syrette immediately; may be repeated after 3 hours, if needed. Do not use morphine within 2 hours of a previous injection, when a person is unconscious, when he has a head injury, or when he breathes less than 12 times a minute.

(2) *Blast concussion injury.*

(*a*) *Cause.* Blast concussion injury is often incurred by swimmers in an area where depth charges, torpedoes, or aerial bombs are exploding. The blast, transmitted through the water, is likely to cause injuries to the lungs, stomach, or intestines.

(*b*) *Prevention.* If expecting blasts described above, float on the back, cross the legs, tense the body, tighten the anus, and keep as near the surface of the water as possible. Get out of the danger area and out of the water as soon as possible.

(*c*) *Symptoms.* If lungs are injured, breathing will be difficult. The patient may spit or cough up frothy blood and may feel abdominal pain. The stomach may be swollen or rigid. Shock may be present.

(*d*) *Treatment.* Lay the victim down with his head low. Keep him warm. If available, give morphine to ease the pain.

(3) *Eye inflammation.*

(*a*) *Cause.* Shipwreck victims are often covered with a heavy coating of dirty oil. The chief danger is eye inflammation.

Wind and sun glare will also cause eye inflammation.

(*b*) *Prevention.* Keep eyes covered with a cool, damp cloth during the day or improvise an eyeshade. Keep eyes closed or above water when swimming in oil-covered water.

(*c*) *Symptoms.* Eyes look oil-stained and dirty. They are red, bloodshot, overflowing with tears, and sometimes painful. Often there is a sticky crust on the lids. Looking at a light causes pain.

(*d*) *Treatment.* Cleanse and apply eye dressing from first-aid kit. To relieve pain, cold compresses can be placed over the eyes 10 minutes out of every hour if there are no ulcers in the eyes.

(4) *Bleeding.* Bleeding must be controlled at once. Application of a pressure bandage will be all that is necessary in the majority of cases. If bleeding persists or the bleeding is from a large artery in the arm or leg, a tourniquet must be applied. The tourniquet can be made from strips of cloth torn from shirts or trousers, handkerchiefs, belts, or other similar material.

(5) *Wounds.*

(*a*) *General.* Stop bleeding by a compress bandage applied to the wound or by applying tourniquet when necessary. Cover the wound with a sterile dressing. Treat for shock.

(*b*) *Chest wounds.* Any wound which penetrates the chest and allows air to enter it may cause collapse of a lung, possibly death. To prevent this, immediately apply a dressing to the wound and make airtight by applying folded pieces of relatively impervious material such as sailcloth or raincoat, and adhesive.

(6) *Fractures of arm or leg.* In case of fracture only, the broken limb can be fixed in position by splints. If the broken bone has penetrated the skin, do not try to push the bone into place. Cut the clothing away from the fracture site, apply bandage, then splint the limb. Handle the limb gently. Maintain a slow steady pull on the limb as the splints are being applied. Care must be taken that bandages do not become too tight either from skrinkage of wet material or from swelling of the limb. Elevate the limb to the most comfortable position.

(7) *Frostbite and freezing.*

(*a*) *Cause.* Insufficient shelter from wind or water or prolonged exposure to cold may cause frostbite and freezing. Frostbite is the freezing of single parts of the body, most often the nose, ears, cheeks, fingers, and toes. If tight clothing reduces circulation, the extremities may freeze.

(*b*) *Prevention.* Adequate, loose, and dry clothes are important. Rig up all available shelter from the wind and water. In temperatures below freezing, wet blankets can be frozen and used as windbreaks. Stimulate circulation by movement. Rest.

Overexertion causes perspiration and loss of energy. Stay out of the water. Keep low in the boat and out of the wind. Keep huddled together. Do not expose the extremities to the wind. Keep facial hair cut short. If tourniquet is applied to stop bleeding, release pressure frequently and warm those parts of the body where circulation has been stopped. Never drink alcohol, as it seriously disturbs the control of body temperature and increases the possibility of frostbite.

(*c*) *Symptoms.* Frostbitten skin becomes white in color; flesh becomes stiff and numb. Symptoms of advanced freezing are muscular weakness, stiffness of limbs, and drowsiness.

(*d*) *Treatment.* Treatment of frozen parts of the body should be started immediately. Change into warm, dry clothing, if possible. Frozen arms, legs, hands, face, etc., should be warmed gradually by placing against some other part of the body which is warm. Do not exercise or rub frozen parts as skin tissue and frozen bones break readily. Never apply snow, ice, kerosene, or oil to the affected parts as this increases freezing. A frostbitten person should not get too near a stove or apply hot water to affected parts, as too rapid thawing causes pain and damages skin tissue. Blisters that appear should be kept clean. If they break, the skin should be trimmed and the sores treated to prevent infection. Do not use strong antiseptics such as iodine. If breathing has stopped, give artificial respiration and warm body gradually.

(8) *Immersion foot.*

(*a*) *Cause.* Immersion of the feet in uncomfortably cold water for several hours or more causes immersion foot. It may be made worse by keeping knees bent to conform with the cramped quarters in the boat.

(*b*) *Prevention.* Keep feet dry and warm. Remove tight shoes. Straighten out legs and elevate feet. Grease the feet and wrap them loosely in cloth to protect against the cold and moisture. Storm oil may be used to oil the feet.

(*c*) *Symptoms.* The first thing noticed is pain in the feet, followed by swelling of the feet and legs. The skin becomes discolored and blood or water blisters or ulcers may develop. The feet feel numb and may become paralyzed.

(*d*) *Treatment.* Do not apply heat, avoid rubbing, and never allow any weight to rest on feet or legs. Raise the legs and feet above the level of the body, being careful not to damage the skin. Keep the rest of the body warm. Apply cold packs to the feet and legs but do not let the skin get wet. Either a cold compress separated from the skin by a layer of waterproof material or cold, dry air blown over the skin is effective. Continue treatment and rest until swelling and pain disappear.

(9) *Burns and sunburn.*

(*a*) *Cause.* Burns may be caused by swimming in burning oil, by an explosion, or by exposure to the sun.

(*b*) *Prevention.* Keep the body completely covered even in cloudy weather.

(*c*) *Symptoms.* The skin is highly red in color, irritated, and usually blistered. A burning sensation is felt. The skin is sensitive to the touch. Fever and shock may be present.

(*d*) *Treatment.* Cover burned area with foille for burns from first-aid kit. Dab, do not rub. Treat for shock. For fever, make patient rest and give him cold fluids to drink, preferably water.

(10) *Heatstroke (sunstroke).*

(*a*) *Cause.* Heatstroke results from exposure to heat and sun.

(*b*) *Prevention.* Retain all clothing and headgear. Rig up awnings from sail, canvas cover, or other material. Take an occasional short swim.

(*c*) *Symptoms.* Symptoms are dizziness, nausea, vomiting, diarrhea, fever, headache, mental confusion and unconsciousness. The skin is red, hot, and dry.

(*d*) *Treatment.* Loosen all clothing. Bathe head, face, wrists, and body in cool sea water. Give small sips of fresh water. Place the patient in a reclining position protected from the sun. Fan and keep cool. Apply cold water to head and extremities.

(11) *Dehydration and thirst.*

(*a*) *Cause.* The body loses water by breathing, by evaporation from the skin, and internally through the kidneys.

(*b*) *Prevention.* See paragraph 41.

(*c*) *Symptoms.* Symptoms are loss of weight, rapid pulse, fever, convulsions, shock, and inability to urinate. Dryness causes cracks and sores on lips.

(*d*) *Treatment.* Give small amounts of sweetened water if the individual is conscious. Treat for shock.

(12) *Starvation.*

(*a*) *Symptoms.* Symptoms are loss of weight, fever, and shock. Breathing may be shallow and fast. Prolonged malnutrition may cause swelling of the feet—not to be confused with immersion foot.

(*b*) *Treatment.* Give small amounts of soft and liquid foods. Keep warm. Treat for shock.

(13) *Constipation.* With little food and water you will have few or no bowel movements. Do not worry about it. Constipation in itself is not harmful in this case. No first-aid treatment is necessary. Do not take laxatives. Laxatives only absorb water from the body and increase the process of dehydration of the body.

(14) *Fainting or unconsciousness.*

(*a*) Lay the patient flat on stomach, head turned to one side.

(*b*) Loosen clothing.

(*c*) If patient is breathing, use ammonia inhalant.

(*d*) If patient is not breathing, use artificial respiration. Use inhalant as soon as breathing starts.

(15) *Mental disturbances.*

(*a*) *Cause.* Usually mental disturbances are caused by severe hardships, prolonged exposure, thirst, starvation, or drinking sea water. Sometimes they develop when rescue seems probable and the victim becomes overexcited and happy.

(*b*) *Symptoms.* Symptoms are irrational thinking, melancholia, a fixed stare, delirium, and convulsions.

(*c*) *Treatment.* Give victim rest, warmth, and quiet. Prevent the man from injuring himself or leaving the boat. Lash him to the boat if necessary.

(16) *Resuscitation.* Aid to breathing may be necessary with patients who have been under water, received concussion shock, or have been overcome by smoke or oil fumes.

(*a*) *Prone pressure method* (fig. 30).

1. Lay the victim on his stomach, one arm overhead, the other folded under the head with fingertips coming to edge of mouth. Turn the face toward the extended arm. Wipe water, mucous, and loose objects out of mouth and pull the tongue forward. Raise hips to drain water from lungs.

2. Extend and spread legs. Kneel astride the thigh on the side to which the head is facing to be able to observe the face. Your knees must be far enough away from the victim's hips so that pressure can be applied to his lower ribs. With your arms straight, place the palms of your hands on the patient's lower ribs so that the little fingers just touch his lowest rib. The thumbs and fingers are in their natural position, and the tips of the fingers are out of sight just around the sides of his chest. The heels of the hands should be placed as far as possible from his backbone without slipping off (fig. 30①).

3. With arms held straight, swing forward slowly so that the weight of the body is gradually brought to bear upon the drowned person (fig. 30②). This procedure should take about 2 seconds or long enough for the count of "one thousand and one, one thousand and two." Do not bend the elbows while giving artificial respiration.

4. Now swing backwards so as to remove all pressure completely and suddenly to the count of "one thousand and three, one thousand and four."

5. Continue without interruption until breathing is restored. If breathing stops again, resume artificial respiration. In changing operators, maintain the cadence.

(*b*) *Seesaw method (Eve's method).* A drowned man unconscious for as many as

① Starting position.
Figure 30. Prone pressure method of artificial respiration.

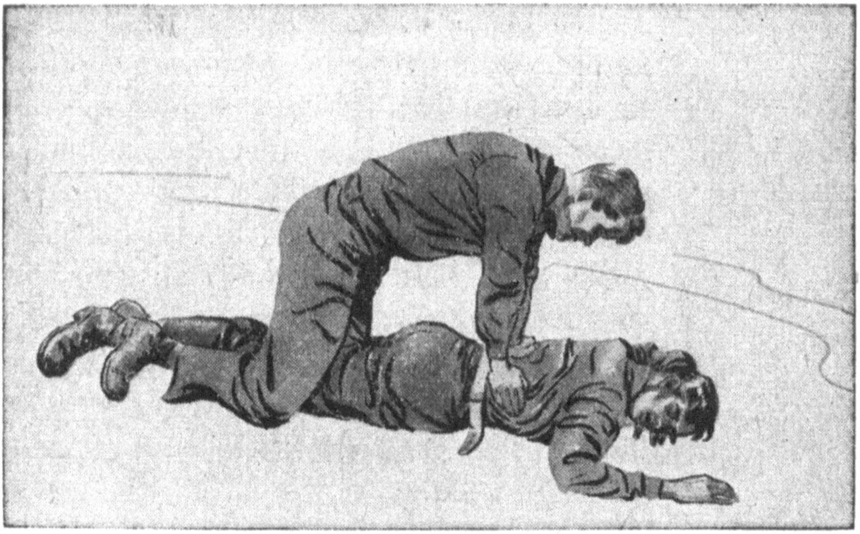

② Forward position.
Figure 30—Continued.

8 hours may be revived by a treatment known as Eve's method, which is based on this principle: If a man's body is tilted with his head downward, the con-

tents of his abdomen slide down toward his head and press on his diaphragm, forcing air out of his lungs; if he is then tilted feet down, the contents of the abdomen slide down toward the feet, pulling the diaphragm down and sucking air into the lungs. Consequently if he is tilted to and fro, his breathing will be done automatically for him. Two or three men are required.

1. Begin prone pressure method at once and continue until a litter and blanket are obtained.

2. Place patient face down on the litter and continue prone pressure method while the victim is secured to the litter. This is done by bandaging wrists and ankles to the handles of the stretcher over heavy padding. An alternate method is to place ropes around his body and the litter just above and below the buttocks. Another expedient consists of a specially prepared board provided with pegs to hold the patient in place and supported by a frame (fig. 31).

3. Meanwhile, if a litter is used, a light rope is secured to a hook, the litter lifted to waist height, and the rope passed below and made fast to a hook opposite. You now have the litter resting with its middle on a loop of rope.

4. Begin tilting the litter or board to and fro and cover patient with warm blankets immediately.

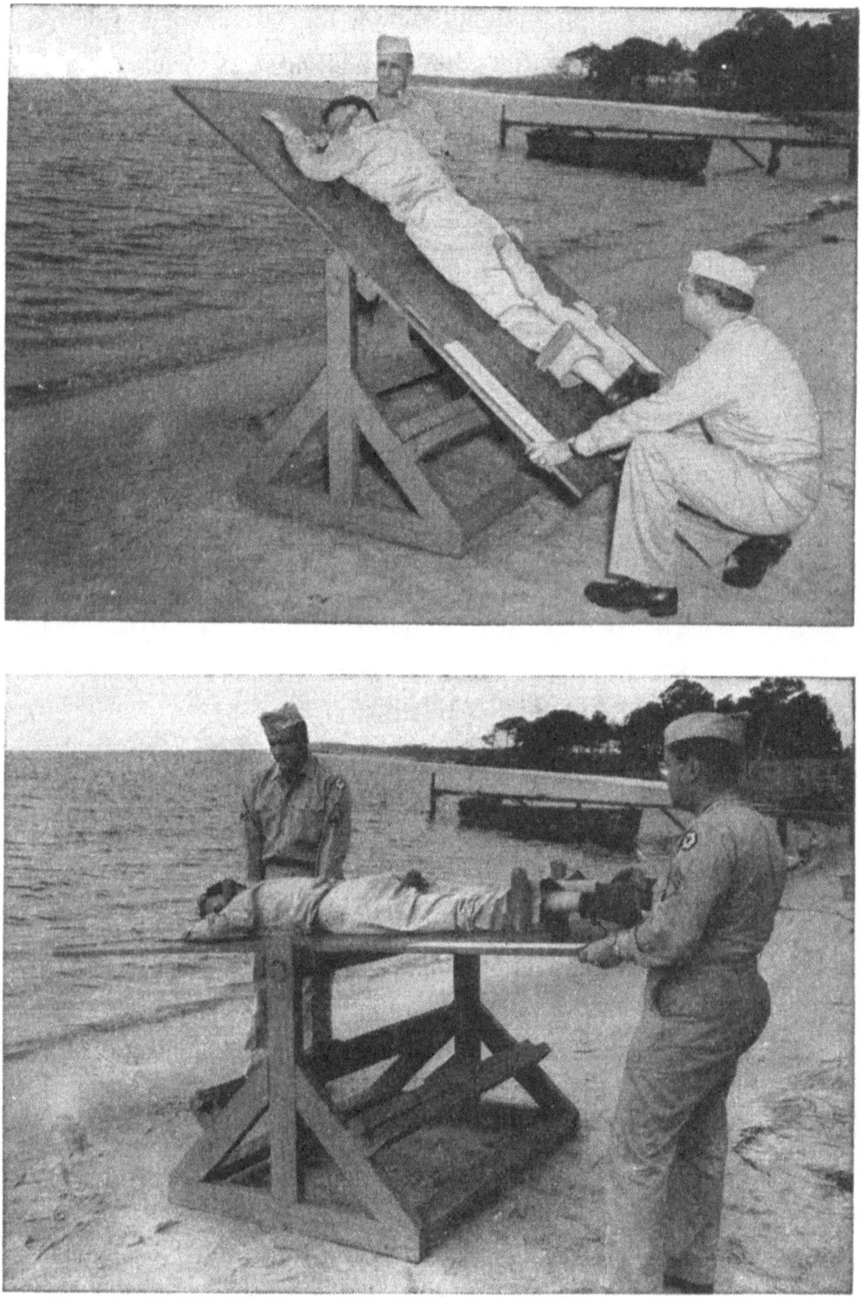

Figure 31. Patient being resuscitated by Eve's method on specially prepared board.

5. Continue the seesaw rocking at the rate of 12 times a minute until normal breathing returns. Rocking must go on for 8 hours if necessary, unless rigor mortis is present or a medical officer declares the victim dead. To prevent the litter from slipping on the rope, nails should be hammered into the under side of the litter bars on each side before the victim is placed on the litter.

(c) *Advantages of Eve's method.* There are many practical advantages to Eve's method. Any untrained man can assist after watching for only a few moments and pick up the rhythm. Thus, it can be kept up for many hours with unskilled help. With several casualties, prone pressure method may be impossible to administer but any novice can rock a litter. The head-down position allows water in the lungs to run out. Wet clothes can be removed and warm blankets put on. When necessary, first aid can be given to wounds and burns. Eve's method cannot do harm such as that sometimes caused by unskilled use of prone pressure.

Section III. PROTECTION AGAINST WATER ANIMALS

50. GENERAL. Some water animals attack man only in self-defense. Others may attack if attracted by blood, shiny objects, and light colors such as that of a man's skin. Don't drag your hands or feet overboard. The best defense against water animals is to look for them and detour around them. Observe under water, on the bottom of shoals, among rocks, and at the surface. Swim slowly and quietly. Keep your clothing on if dangerous fish are known to infest the water. Move away from any blood in the water. It is important to remain calm, especially when stung by water animals. Their stings will wear away. Do not worry about whales. The chances of their harming you are rare. Metal struck against metal under water will frighten them away.

51. RECOGNITION OF WATER ANIMALS AND DEFENSES AGAINST THEM. **a. Sharks** (fig. 32). Sharks are distributed widely but are most common in warm seas. They have long, round, slender bodies with the upper lobe of the tail-fin longest, and with five distinct openings to the gills along the side of the head. The most dangerous sharks have unsymmetrical tails. The body normally measures not more than 11 feet. The mouth is large and armed with cutting teeth; the nose usually is conical, bluntly pointed, and protrudes well in front of the mouth. Sharks usually attack on the surface and are revealed by their fins, which break the water like a periscope. Defense against a shark attack consists in splashing, moving the arms and legs rapidly, and making a great deal of commotion under water. Metallic noises such as striking a canteen under water are best. Avoid display of skin, underclothing, or shiny objects. By striking a blow on the snout, the most sensitive part of the shark's body, you may drive him away. Treat in the ordinary manner any wounds received.

b. Barracuda (fig. 33). The barracuda is found only in warm seas. It is a grayish, pikelike fish about 6 feet long, with long, pointed jaws lined with sharp

Figure 32. White shark—"maneater."

teeth. The barracuda usually is attracted by anything that moves but especially by light-colored or shiny objects. It attacks quickly from below the surface. Creating noise under water may frighten it away. Treat any wounds received in the ordinary manner.

c. Sting ray (fig. 34). Sting rays, found in all warm seas and in some fresh water rivers, are disk-

Figure 33. Barracuda.

shaped, flattened fishes with one or two long, barbed spines. The disk is made up of both body and fins. Sizes range from that of an ordinary dinner plate up to 10 feet across. The tail may be as long as or longer than the disk. Since they conceal themselves in muddy or sandy flats, sting rays are frequently stepped upon and lash out with their tails, driving the spine into the flesh and injecting a highly venomous substance. Treat wounds like snake bites. When walking in turbid or muddy water, poke a stick ahead of you and slide your feet along. If stick or feet touch a hidden sting ray it will swim away.

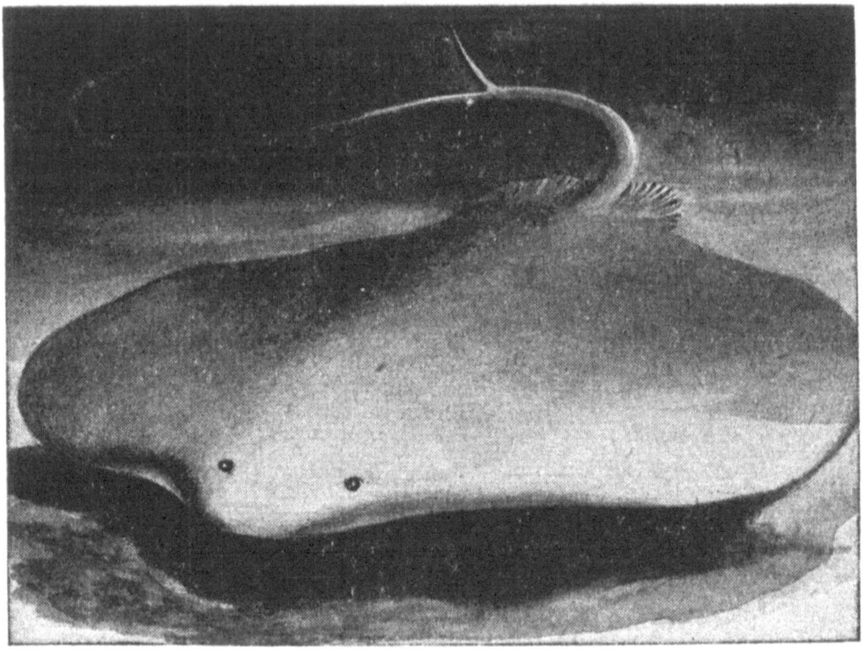

Figure 34. Sting ray.

d. Sawfish. The sawfish has a body similar to the shark's. In addition, a swordlike snout is armed with spines on the sides giving the impression of a large double-edged saw. The sawfish may reach a length of from 10 to 20 feet but it is not vicious. It lives over sandy and muddy bottoms. The sawfish can swing its saw back and forth with power enough to break a man's leg. The most tender parts of the sawfish are the areas around the eyes and at the base of the saw. Defense consists of striking area of eyes and avoiding the saw.

e. Moray eel (fig. 35). Moray eels are found in all warm seas, especially in crevices about coral reefs. Most morays are brownish or blackish colored with peculiar patterns of varied spots. Some morays reach a length of 6 feet. A knife or spear may be used in defense against the moray. Keep your hands and bare feet out of rock crevices. Treat in the ordinary manner any wounds received.

Figure 35. Moray eel.

f. Sea porcupine (fig. 36). Found in warm seas, the sea porcupine is recognized easily by its covering of erectile spines. This fish swells itself by swallowing water or air. It is not ferocious. Its jaws are like the beak of a parrot and powerful enough to bite off a finger if it is molested. Defense consists of keeping out of its way.

g. Sea urchin (fig. 37). The sea urchin occurs abundantly on rocks, reefs, and among coral and looks like a pincushion full of long needles. Sea

Figure 36. Sea porcupine.

urchins with short, stout spines are not poisonous. Some sea urchins are covered with numerous movable spines of two different sizes. The shorter and finer spine is highly venomous; if it touches the skin gently, the poison is injected into the flesh causing sharp, severe pain. Remove the spine and apply iodine. Be suspicious of anything that resembles a sea urchin; do not handle it.

h. Giant clam. Found on the coral reefs in the Pacific and Indian Oceans, giant clams are similar to the ordinary clam but of tremendous size, sometimes weighing more than 500 pounds. The clam is edible, but care must be taken that no part of the body is trapped within the shell, which clamps on to anything that enters it.

i. Octopus (fig. 38). Commonly found in the Mediterranean and the Southwest Pacific, the octopus

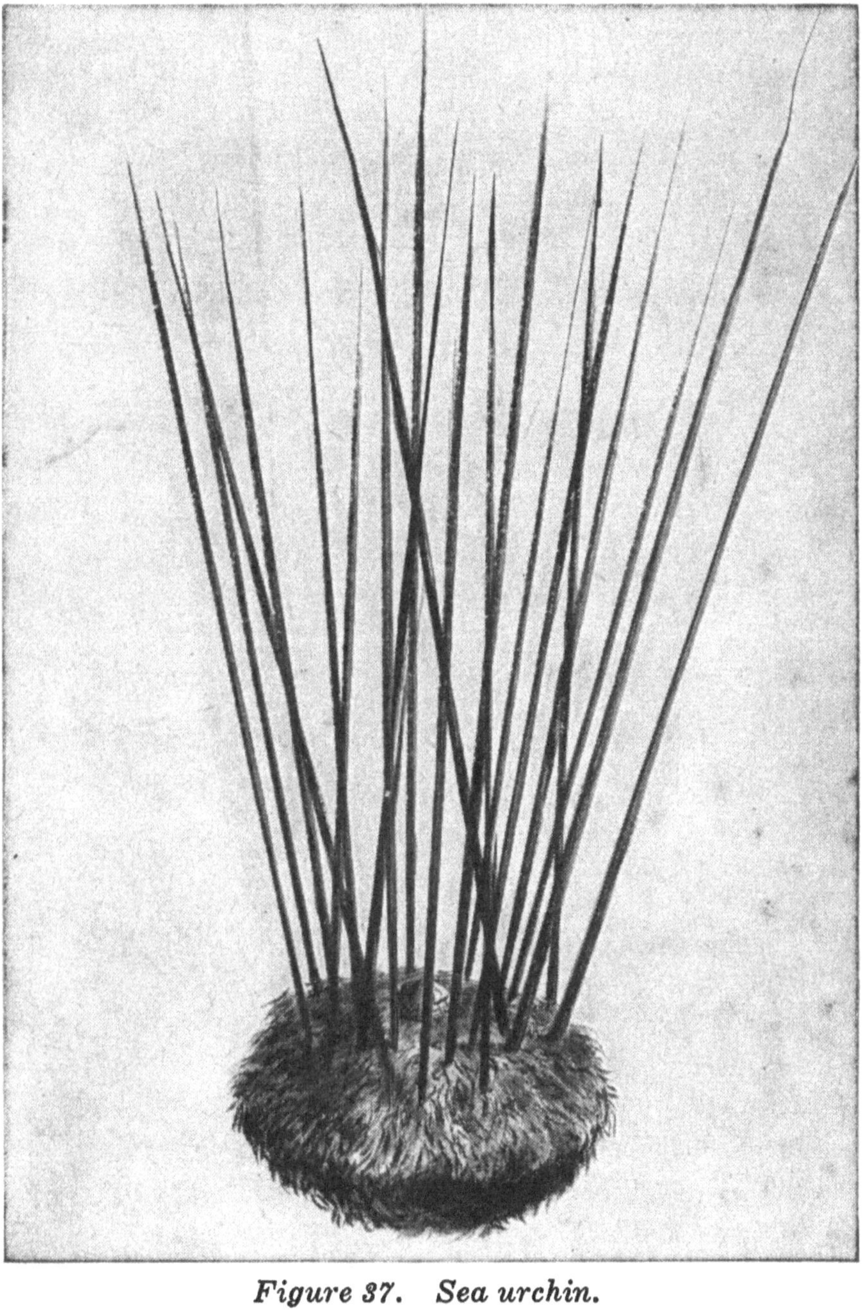

Figure 37. Sea urchin.

has a round body and eight arms or legs on each of which are numerous suction cups. The octopus may grow to over 10 feet from tip to tip of the tentacles. It has large keen eyes which shine in the dark. When attacked, it emits an inky fluid into the water to screen its actions while escaping. The octopus is not a vicious animal and when attacked will try to escape. It occurs most frequently along rocky shores and on reefs. Some kinds live at considerable depths in the sea.

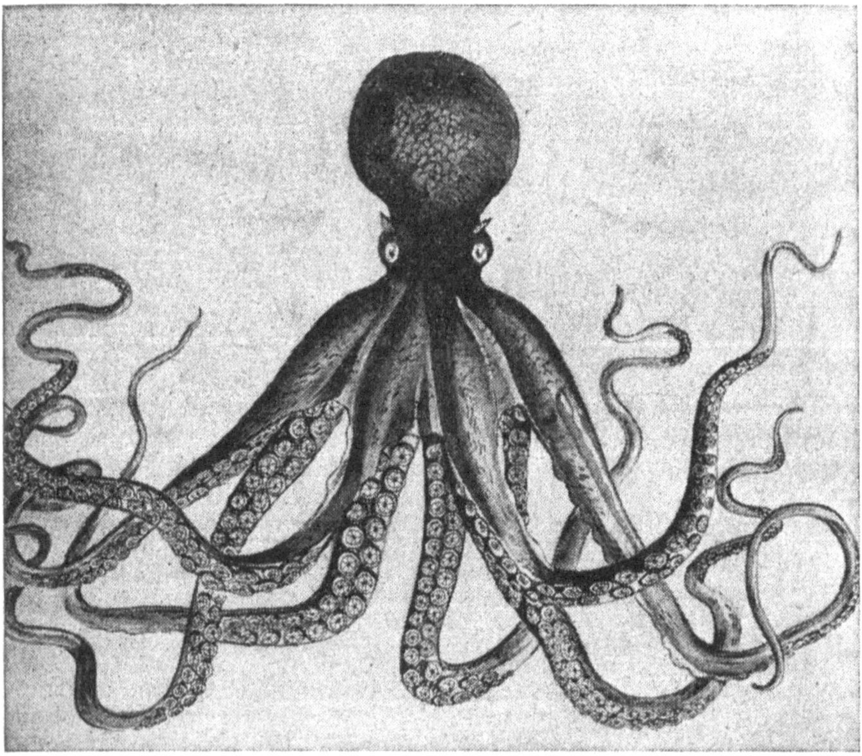

Figure 38. Octopus.

j. Electric ray. In shape, the electric ray is similar to the sting ray except that its tail lacks the sting. On being touched, the animal imparts a severe electric shock from batteries located along its back. If shocked, remain calm and quiet and wait for the shock to wear off.

k. Bluefish. Bluefish are unusually active game fish abundant in the Atlantic Ocean and English Channel. They are generally 2 to 3 feet long and blue in color. They have razor-sharp teeth and will attack any moving object. Schools of bluefish are dangerous to a swimmer. Treat inflicted wounds in the usual manner.

l. Jellyfish. Found in all seas but more numerous in the Tropics, the jellyfish is an umbrella-shaped animal of jellylike substance, with numerous tentacles hanging down from the under side. Jellyfish vary from a few inches to 2 or 3 feet across. Contact with a tentacle causes a severe stinging sensation. Application of slightly diluted ammonia water gives immediate relief. If stung while swimming, remain calm and swim slowly until the effects wear off. The jellyfish cannot follow you. Clothes give protection.

m. Portuguese man-of-war (fig. 39). Commonly found in most seas, the brightly colored Portuguese man-of-war has a large bladderlike body with long tentacles hanging down from the under side. It usually floats on the surface. It imparts a more severe sting than a jellyfish, but the sting is treated in the same manner. Watch for the floating bladder and keep away from it.

n. Sea snake (fig. 40). Sea snakes can be distinguished from eels because they are covered with bony plates or rectangular-shaped scales. They are found in the warm waters of the Indian and Pacific Oceans and, a fresh water variety, in the Philippine Islands. The sea snake is usually banded with bright

Figure 39. Portuguese man-of-war.

colors. The tail is flattened to form a paddle. Sea snakes rarely bite without provocation, but stay away from them. Their venom is poisonous; treat wound immediately. First apply a tourniquet between the bite and the heart. Next with a knife make a criss-cross out of each fang prick; suck out the blood and poison by mouth and spit. Finally apply iodine. If your mouth contains any open wounds, have someone else suck the poison.

Figure 40. Sea snake.

o. Crocodile and alligator. Crocodiles and alligators are found in fresh water in Africa, Asia, Australia, and America. However, the largest and most dangerous crocodiles take to the open sea in the Indo-Australian region. Crocodiles and alligators are long, thick-skined reptiles with a vicious, lashing tail and a long snout with big teeth. Stay away from them.

CHAPTER 7

RESCUE

Section I. ATTRACTING ATTENTION

52. GENERAL. Do not waste signaling equipment on the chance someone may see your signal. A real chance of being rescued may be lost a few hours later. If an airplane is heard in your vicinity, wait until it is heading in your direction and fairly close to you before firing a signal. Remember, you can hear an airplane long before you can see it or the airplane's crew can spot you. Be prepared to fire a second signal to confirm the first. Be sure you are signaling a friend, not an enemy. In the absence of signaling equipment, make yourself conspicuous by churning up the sea with oars or paddles.

53. RADIO. Lifeboats may be equipped with portable emergency radio transmitters. Instructions for sending signals are printed on the set and are also contained in the tube holding the antenna kite and balloons. Motor lifeboats usually have permanently installed radios. Figure 41 illustrates one method of erecting an antenna.

a. Precautions. If you have been torpedoed, do not use the radio until you are reasonably certain the enemy craft has left the scene of action. The "Gibson Girl," one of the most common types of portable emergency radio, has a sending radius of 250 miles, and signals have been picked up as far as 400 miles away. Do not send signals, therefore, when you are

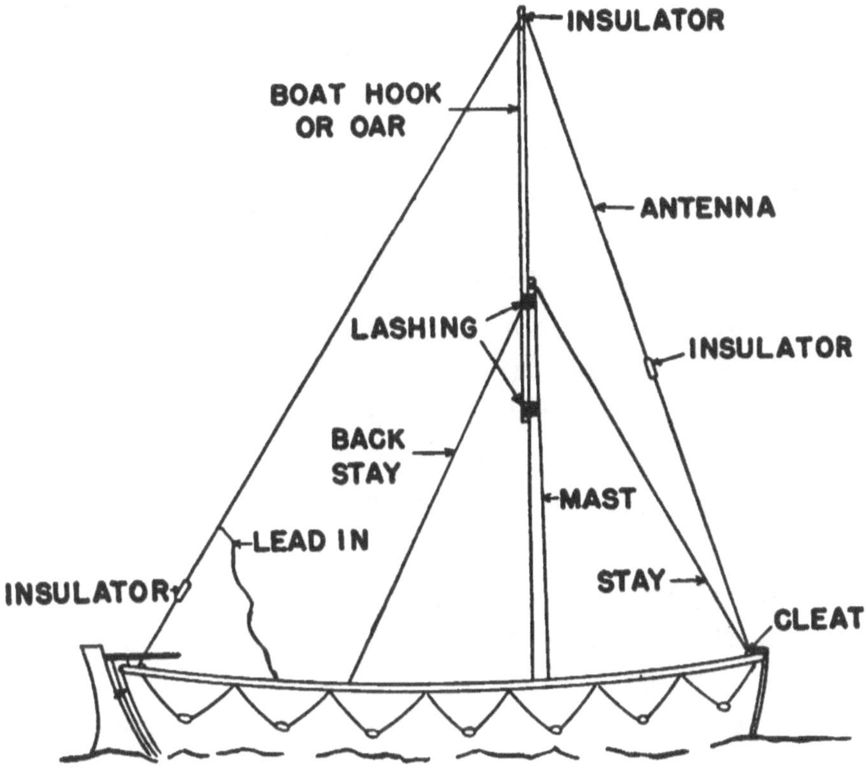

Figure 41. Method of rigging radio antenna.

within 250 miles of enemy territory. Otherwise, get on the air as soon as possible after being set adrift. Do not let the antenna sag in the water, but be certain the ground wire is in the water. Take down the antenna during storms or when there is lightning. Use the hydrogen balloon instead of the kite when the wind is less than 7 miles per hour. When using the hydrogen generator, do not smoke and be careful not to spill any of the contents, as the chemicals may burn the skin and injure the craft if you are in a rubber boat.

b. Sending signals. Many emergency radios are equipped with an automatic device for sending out an SOS, as well as with the usual manually operated key. Send signals during the 3-minute international silent periods, which start at 15 minutes and 45 minutes after the hour, Greenwich civil time. Send signals when a friendly airplane or vessel is heard or sighted. Crank steadily so rescuers can take accurate bearings.

54. SIGNALING MIRROR. a. Description. A double-faced mirror with about 20 square inches of reflecting surface on each side and a sighting hole in the center is provided in most lifesaving craft. The mirror is coated with grease and wrapped in a marked container. In clear weather, flashes may be seen up to 10 miles. As a substitute for a signaling mirror, an ordinary pocket mirror or any other reflecting surface, such as a flattened tin can or the wet blade of an oar, may be used.

b. Use (fig. 42). To use a signaling mirror, follow this procedure:

(1) Face a point about halfway between sun and observed object.

(2) Hold mirror in one hand about 3 inches from the face and sight the object to be signaled through the sighting hole in the mirror.

(3) The light from the sun shining through the hole in the mirror will form a spot of light on your face which will be reflected on the side of the mirror facing you. Now, while still sighting the object, adjust the angle of the mirror so that the spot of light reflected on the mirror disappears in the hole in the mirror. When this occurs, you will know that the reflected light from the sun is being directed toward the object.

Figure 42. Use of signaling mirror.

55. OTHER SIGNALING DEVICES. a. Lights and flares. Instructions for the use of signal pistols, flares, smoke signals, and distress lights, all of which are normal lifeboat equipment (par. 5*c*), are found in the watertight containers holding this equipment. The lantern is a valuable night light as a check on the watch, should rescuers approach the vicinity undetected. A flashlight can also be used to signal at night and can cover wide areas and serve as a last resort, when all other night signaling equipment has been expended. You can signal an SOS with it by three short flashes, three long, then three short.

b. Signal flag. The best method of displaying the signal flag is for two men to stretch it taut by holding each end, then move it from side to side (or from vertical to horizontal if the rescuer is in the air) in order to present a flash of color to attract attention. Triced high to the mast, the signal flag can best be seen by lookouts on passing vessels far on the horizon.

c. Boat cover. Whether or not you are using the tarpaulin or boat cover as a canopy, display it with the painted side up. It may be waved when a rescue craft is sighted.

d. Whistle. At night or in fog, use the whistle to attract surface vessels or people ashore, or to locate another lifesaving craft if it becomes separated.

56. OTHER PROCEDURES. a. Protection of equipment. Protect your signal equipment as much as your rations. Place a responsible man in charge of it.

b. Attracting attention at night. To aid in attracting attention at night, churn or splash the sea with oars, paddles, or the bare hand. This will reveal more clearly luminous particles in the water.

c. Assembly. Unless the sea is very rough, shorten the line between rafts or boats if you see or hear a friendly plane. Two or more lifesaving craft together are easier to spot than the same number widely separated.

Section II. RESCUE AT SEA

57. MAINTAINING ORDER. Order must be maintained and navigation must continue without interruption. Do not change your course until land has been positively sighted. Your boat may drift away from land before it can be beached, or the surf may be so heavy you cannot get ashore. The rescue craft may not see you, or it may turn out what you see are actually other survivors or enemy craft. Remember the following points:

a. Don't change your course to reach a sighted craft. If it is a friendly rescue craft, it will approach you and probably be in better shape to do so.

b. Don't dispose of any food or water even if it appears that rescue is possible.

c. Don't consider yourself rescued until you are actually aboard the other craft and under the orders of her commander.

58. RESCUE PROCEDURE. Once contact has been made with rescuers, follow their directions. After sighting you, rescuers may choose to wait for daylight or calmer seas before picking you up, according to the weather, condition of the survivors, and the ability to maintain contact.

a. Precautions. To expedite your rescue after being sighted, follow these instructions:

(1) *Maintain identification.* If you use a mirror, it will enable rescue craft to keep trace of your position.

(2) *Signal if injured badly.* You can do this by crossing the arms across the body, or by improvising a cross out of material at hand.

Two colored life-raft sails may also be used. If uninjured, hold arms straight out at side. A surface craft may acknowledge your signal with a blast of the whistle, or a plane by dipping wings.

(3) *Do not use emergency equipment not required.* It may not be possible to pick you up at once, so conserve your gear, particularly distress lights, which are of little value by day.

b. Methods of rescue. You will be picked up by a flying boat, a seaplane, submarine, or surface ship. If initial contact is made by a plane of a rescue squadron, detailed instructions regarding cooperation expected of you for each type of rescue will be dropped to you. Remember these points:

(1) If you are being rescued by a flying boat, continue to signal by mirror. Stay clear of propellers; a swell can easily wash you into them. A life preserver with a line will be thrown to you, or, if you are incapacitated, a man in a raft will assist you to the plane. If the sea is unfavorable, the plane will report your position and arrange for you to be picked up by a submarine or surface craft. This plane will drop provisions as required for your rescue.

(2) If being rescued by a seaplane, continue to signal when the plane is on the water. Do not ride a lifeboat close to the plane or hold on to the wingtip float, as there is danger of capsizing the plane. Swim or paddle to the main float aft the wing. Follow instructions

from the plane's crew. A life preserver and line may be thrown you.

(3) If being rescued by a submarine, continue to signal with mirror. If the submarine surfaces, look for life preserver and line to be thrown. A periscope tow may be desirable in wartime if shore batteries are close. A line that can be slipped quickly should be secured to the periscope in case the submarine dives. Best tow length is 20 feet behind the periscope. Maintain contact through periscope and keep watch for unfriendly craft. If rescue is by night, answer flares from submarine with flares or flashlight. Listen for whistle; answer with whistle, flares, or flashlight.

(4) If being rescued by surface ship, signal by day with mirror; by night, the best signals are flares, then flashlight.

59. DISTRESS SIGNALS. The following distress signals have been adopted by the U.S. Coast Guard.

a. Red light or red rocket or flare at night: "You are seen; assistance will be given as soon as possible."

b. Red flag waved ashore by day, or a red light, red rocket, or red roman candle displayed by night: "Haul away."

c. White flag waved ashore by day, or a white light swung slowly back and forth, a white rocket, or white roman candle fired by night: "Slack away."

d. Two flags, a white and red, waved at the same time ashore by day, or two lights, a white and a red, swung slowly at the same time, or a blue light burned by night: "Do not attempt to land in your own boats; it is impossible."

e. A man ashore beckoning by day, or two torches burning close together by night: "This is the best place to land."

f. Any of these signals may be answered from the vessel as follows: In the daytime, waving a flag, a handkerchief, a hat, or even the hand; at night, by firing a rocket, a blue light, or a gun, or by showing a light over the ship's gunwale for a short time, and then concealing it.

Section III. REACHING LAND

60. LANDING AND BEACHING. Do not let the sight of land build your hopes up too high. Some land has less to offer than the open sea. When you are sure you have sighted land, and fog or night starts to set in, take a compass bearing and then steer by the compass. Do not land in surf, rough sea, or at night if it can be avoided. Besides the natural hazards of landing at night, you run the added risk in wartime of not hearing a sentry's challenge and of encountering land mines and underwater wire entanglements. If there are signs of life ashore, it will be wise to remain outside the breakers and try to signal the shore for help for someone with knowledge of local conditions and with a larger craft. If you have to wait overnight before landing, you should not wait too close to the shore or the reef lying out from the shore. At the same time, take care not to drift out of sight of land during the night, especially in rainy and windy weather. Try to find a relatively calm place to land. Often it is necessary to search for some time before finding a suitable beach. Because of the formation of bars, shore line, or rock there will be certain spots along the beach where the surf will be easier to run. It may be difficult to tell how rough the surf is when approaching a beach because the surf appears smoother from the seaward side than it really is. The outer line of breakers is ordinarily the heaviest so that a boat taken through them will have a good chance of reaching shore safely. An inlet may appear difficult because of high surf breaking on an outer bar, but it may be discovered that there is a pass between the extremities of the bar and the beach proper. By going to the leeward end of the bar, it may be possible to pass in smooth waters without entering the breakers at any time. Passages in reefs are indicated by calm gaps in the long line of breakers. If the pass is deep, the color will be clear blue; if shallow, it will be a blotchy brown. Off-lying rocks may smooth out the breakers and provide a clear channel. However, such beaches must be approached with care because of the danger of submerged rocks. Adjust life preserver and put on shoes before making an attempt to land.

a. Approaching a beach. Wait outside the breakers and study the seas as they roll in. Usually there will be a succession of heavy swells, then a period of relatively smooth water. An outstanding hazard is the possibility of turning sideways. This can be avoided if the boat is kept end-on to the breakers by rowing against the sea and using a drag or sea anchor. This will also help reduce the speed of the approach and give more control over the boat. In all instances the boat should be steered by an oar over the stern or on one quarter. Sails and masts should be taken down outside the breakers. Heavy weights should be kept out of the extreme ends of the boat. It is best to keep the seaward end heavier than the shoreward end to help prevent turning sideways. Storm oil may be used to reduce the effects of the breakers. There are different methods of landing and beaching and the situation will govern which to use. Two methods are as follows:

(1) *Landing stern first.* Before entering broken water, turn the bow of the boat to the sea. Do this where the surf is lighter, as in the relative lee and quiet of a point or bulge in the shore line. Such a spot will probably be found where there is a gully or break in the shore line or a row of cliffs behind the beach. Back the boat in stern first. When a breaker approaches from seaward, the crew should pull a few strokes to meet it and then resume backing in towards shore. Landing stern first is considered the safest procedure with an inexperienced crew.

(2) *Landing bow first.* Row to shore, bow first, by backwatering on the approach of a wave and rowing ahead again when it has passed. A sea anchor or a makeshift drag weighted down by an anchor can be used to prevent the boat from turning sideways. This method is used with an experienced crew and a relatively calm sea.

b. Beaching a lifeboat. Whether a lifeboat is brought in bow end first or backed in (stern first), she is kept at right angles to the shore when beaching. When the boat reaches shore, the crew will jump out, grab the sides, and pull her in.

c. Beaching a power lifeboat. A power lifeboat should enter the surf at a moderate speed with the rudder unshipped. A steering oar and an oar out on each quarter is lashed in place to assist in steering. If the surf is dangerous and breaking close to the beach, it is safest to stop the engine and land under oars. Approach to shore should be deliberate and all effort directed toward keeping the stern aimed directly at overtaking seas. To obtain this slow approach and certainty of aim, throw astern a sea anchor or makeshift drag weighted down by the anchor to give firmest hold on the water. This use of the sea anchor will check headway of the boat and hold her stern directly into the crest of overtaking seas. It is the foremost protection against a sea breaking into and overturning the lifeboat. In addition to the use of a sea anchor, check forward motion by reducing or reversing the engine or backwater with the oars. If circumstances warrant, have a man ready to cut the tripping line and the sea-anchor rope. Keep a strain on the rope because slack rope may foul the propeller. Because it will change the course of the boat, reversing the engine in a surf is dangerous and should be done only to check the forward motion of the boat. Weight should be distributed to trim the boat by the stern, causing it to drag.

61. AFTER REACHING SHORE. If you happen to land in an uninhabited area, your main needs will continue to be water, food, and shelter. If you can find water and food, it may be advisable to rest a little, regain strength, and then start out again in your craft. In any case, be prepared to make appropriate signals to attract attention should a rescuer approach the vicinity.

a. Obtaining water.

(1) *Ground water.* Do not run any risks from ground water. No matter how thirsty you are, take the trouble to boil it for 1 minute or purify it with halazone pills or iodine from your first-aid kit. The halazone must stand from a half hour to an hour and the iodine an hour or two, but it is worth it. Stagnant water is dangerous anywhere in the tropics. When purifying water with halazone, use two halazone tablets to each canteen of clear water or four tablets if water is cloudy, insert stopper, and shake. After a half hour, shake the container and remove stopper.

(2) *Rain and plant water.* Collect rain in the same manner as at sea. A lot of water collects in the large, lower leaves of trees. You can make containers of large leaves or of bamboo, or dig a hole and line it with a piece of heavy cloth. Big ropy vines or lianas that hang down from trees in the Tropics also furnish water.

(3) *Digging for water.* Along ocean beaches and near them you can dig for water. Dig a hole at low tide just below the high-water mark. Do not dig deeper than the first water you find. Fresh water is lighter than sea water and will stay on top of it. Skim the water off with a leaf, spoon, or other materials at hand.

(4) *Ice and snow.* Eating snow will have no

adverse effects if you eat a little at a time and melt it in your mouth before swallowing. If fuel is scarce, place snow and ice on a dark surface, such as a rock, where the sun will melt it. Catch the drippings in a container.

b. Obtaining food.

(1) *Fish.*

(*a*) *Sea fish.* Fish of many kinds are usually plentiful on reefs, in lagoons, along the shore, and just offshore. You can use the usual methods of hook-and-line fishing, spear them, hit them with sticks, or trap them in improvised nets or in pools among rocks and coral blocks. A few shore fish are poisonous (par. 42*b*(4)).

(*b*) *Fresh water fish.* All fresh water fish, including crustaceans such as crabs and crawfish, must be cooked before eating. To cook the crustaceans, drop them into boiling water. Use a dip net to catch fresh water shrimps. They often cling to branches that hang in the water. They can also be caught by building a dam in a stream. All fresh water or land crabs and lobsters, as well as the sea variety, are good to eat if fresh. All land crabs must be cooked because they are often infected. Do not eat the eggs or liver of any fish.

(2) *Animals and insects.* Turtles and turtle eggs are a good food. Turtles come ashore, often at night, and dig a hole in the sand in which to lay their eggs. Look for small parallel tracks along the shore. Follow the

tracks to the end and there dig for turtle eggs. Monkeys, rats, flying squirrels, anteaters, bats, lizards, frogs and snakes (even if poisonous, provided they have not bitten themselves) are all good to eat. Grubs found in the ground or in rotten wood, as well as termites, locusts, crickets, and grasshoppers whose wings, legs, and other hard parts have been removed, make good food. Do not eat caterpillars. All birds are good to eat, raw or cooked. In the colder climates seals, walrus, bears, wolves, foxes, hares, etc. may be shot, trapped, or harpooned. Skin immediately, leaving fat with meat, and remove entrails. Do not eat the liver of a polar bear as it is poisonous.

(3) *Plants.* Most tropical fruits are good to eat. At the top of many palm trees is a large tender bud which can be eaten cooked or raw. Rattans long slender vines with curved thorns, have a similar bud. Coconuts contain delicious milk and white meat. Breadfruit, oval in shape with a wartlike surface, can be roasted by putting it in the ground, covering with leaves, laying hot stones around it, and covering with dirt. The durian is a large fruit with spines and an obnoxious odor, but it tastes like custard. Eat it raw. There are limitless other edible fruits, including bananas and plantains, papayas, guavas, mangoes, figs, and berries. Nuts of all descriptions and underground vegetables like yams, taro, cassava, and arrowroot are good to eat. In Arctic regions lichens, ber-

ries, roots, plant leaves, and fungi are edible. Water hemlock and some species of mushrooms and toadstools are poisonous. The water hemlock belongs to the parsley or carrot family and has feathery toothed leaves, small flowers arranged in clusters, and parsniplike roots. The leaves are streaked with purple and have a disagreeable odor when crushed.

c. Obtaining shelter.

(1) In cold country, selection of a site for even a temporary camp for the night deserves careful thought. Dryness is the first requirement. You can improvise a tent from canvas or a boat cover. If you pitch your tent where snow may drift, the tent opening should be sidewise to the wind to prevent its being blocked by snow. A windbreak made of snow will help. Sometimes caves, trenches, or tunnels dug in the snow make good shelters. In wooded areas a lean-to shelter is easy to construct (fig. 43). Avoid camping under overhanging shelves of snow, actively building snow drifts, and at the base of slopes where avalanches would be likely.

(2) In the Tropics do not camp near fresh water because of the presence of mosquitoes and of wild animals which come to drink there. Anopheles mosquitoes range within a mile of their breeding places. Also, in malarious areas do not camp near natives. In warm climates you can make a temporary shelter quickly by dropping a tarpaulin or boat cover over a pole. Use leaves to cover your

lean-to, if you decide to make one, and to make a mat or bed.

d. Obtaining fuel. In some cold regions wood is scarce. Dry plants, moss, peat, and animal dung may be used. If these fuels are not available, bones or stones can be used to support pieces of fat which will melt and burn. Build fire on green logs or rocks so melting snow will not put it out. Fuel may be conserved by building a fire in a hole in the ground or by building a fireplace of stones. The stones can then be placed in a metal container and used to heat the shelter. After a campfire has dried and scorched the ground, the coals may be scattered and trampled: then the ground can be spread with green boughs and bedding.

① Bough shelter.

Figure 43. Construction of shelters.

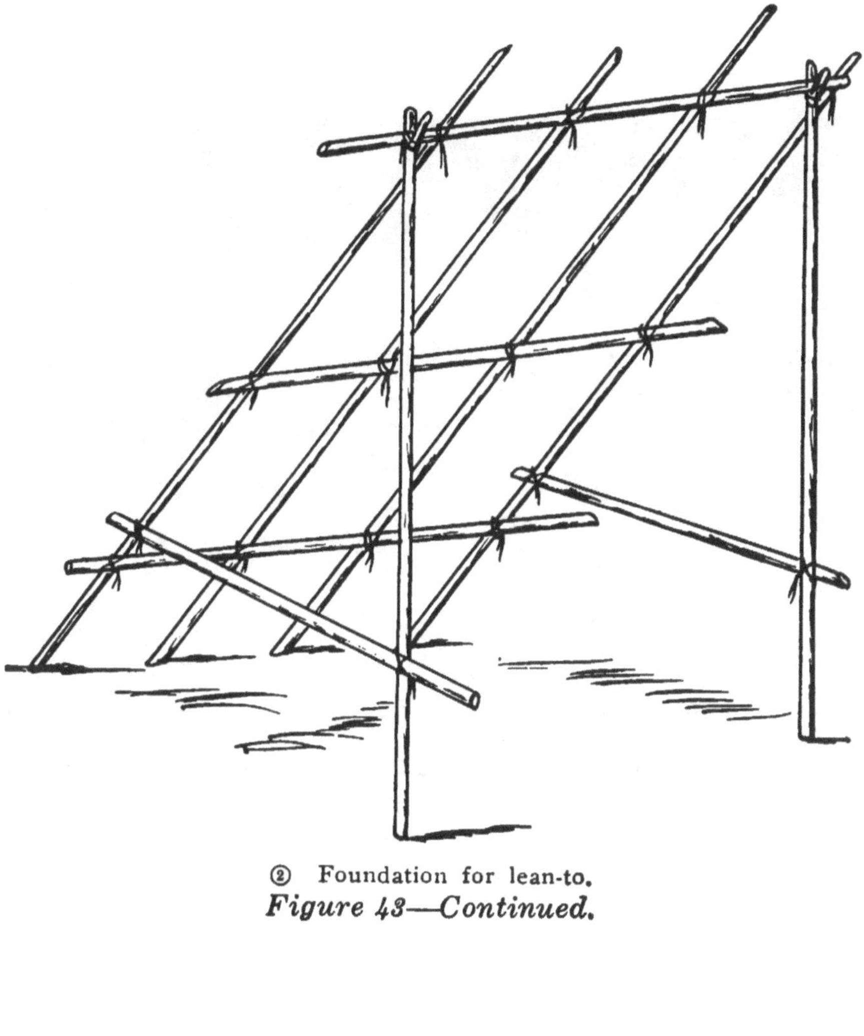

② Foundation for lean-to.
Figure 43—Continued.

③ Three-man lean-to.

Figure 43—Continued.

CHAPTER 8

SUGGESTIONS FOR INSTRUCTORS

62. GENERAL. Instruction in survival at sea should be given at the staging area before embarkation. The instructor should bear in mind and impress upon the students that lack of knowledge of the fundamentals of survival at sea may be the difference between life and death. All instruction should be given in the staging area if time permits; however, if not, give that instruction which requires extensive training aids and practical exercise, and reserve the lectures for instruction aboard ship.

63. TRAINING BEFORE EMBARKATION. Instruction in military swimming, the use of safety equipment, and abandoning ship should be given before embarkation, as extensive training aids and practical exercises are required. Give the men the opportunity to use all safety equipment.

a. Swimming. For men who know how to swim, practice should be conducted with clothing and equipment on. Men who do not know how to swim should be taught basic water skills before being required to wear clothing and handle equipment.

b. Safety equipment. This instruction should be broken down into two parts: individual safety equipment and ship's safety equipment.

(1) *Individual safety equipment.*

(*a*) Explain the use and care of various types of life preservers and the uses of the life-preserver light, whistle, and jackknife.

(*b*) Explain the value of the personal emergency kit, which should be collected and packed under supervision in the staging area.

(2) *Ship's safety equipment.* Actual equipment should be shown and explained. It is recommended that the men be conducted through a vessel provided with safety equipment, if possible; if not, all safety equipment should be laid out as it would be found on shipboard.

c. Abandoning ship. Conduct to be followed and reasons for such conduct should be explained.

(1) *Going over the side.* Explain and demonstrate methods of going over the side. Give the men practical exercise in the use of ropes and nets, and in actual jumping, using safety equipment that they would have under actual conditions.

(2) *In the water.*

(*a*) Explain procedure to be followed when lifesaving craft are not available.

(*b*) Teach procedure to be followed when in lifesaving craft. Instruction and practical exercises should be given, bearing in mind the following points:

1. Command and why it is important.

2. Operation of the lifesaving craft.

3. Use of all equipment in the lifesaving craft.

4. Determination of position and direction.

5. Procedure when rescue is imminent.

64. SHIPBOARD TRAINING. Instruction in first aid; conduct aboard ship; marine, animal, and vegetable life likely to be encountered; and signs indicating land can be taken up aboard ship, if necessary. Training received while in the staging area should not be forgotten and should be reviewed while aboard ship.

a. First aid. First-aid classes can be held in small groups and restricted areas. Demonstration is valuable in these classes.

b. Conduct aboard ship. Individuals should be familiar with the stations they are to man during emergencies, location of the ship's various activities, and the shortest routes to these stations and activities.

c. Food. The men should be familiarized with the marine life in the area and the animal and vegetable life with which they may come in contact. Particular attention should be paid to dangerous animals and to animals and vegetation which are edible.

d. Indications of land. The men should be familiarized with the various signs which may indicate the proximity of land.

APPENDIX I

REFERENCES

FM 21-11 .... First Aid for Soldiers.
FM 21-20 .... Physical Training.

APPENDIX II

GLOSSARY

AFT .... Toward the stern. Between the stern and the midship section of a vessel.

ATHWARTSHIP .... At right angles to the centerline of a ship or boat; across the craft from port to starboard.

BILGE .... Rounded side of ship where it curves up from flat bottom plates to vertical shell plating.

BILGE PUMP... A pump for drawing off water in the bilge.

BLADE .... Flat, paddlelike part of oar on opposite end of handle.

BLOCK .... A metal or wooden frame or shell containing one or more pulleys or sheaves.

BOAT .... Small vessel for oars or sails.

BOAT FALL.... See FALL.

BOATHOOK .... A wooden staff with a metal hook at one end used for fending off or holding on.

BOOM .... A spar used in handling cargo, or as the lower piece of a fore and aft sail.

BOW .... Forward part of a ship or boat.

BREAKER ..... Small cask for fresh water carried in lifeboats; surf.

BUOYANT .....Having the power or tendency to float or keep afloat.

CAPSIZE ......Overturn.

CARBON DIOXIDE RUBBER BOAT ....Collapsible rubber boat inflated by carbon dioxide which is contained in a large cartridge and released with a hand valve.

CARGO NET....See LIFE NET.

CENTERLINE ..An imaginary straight line running the length of a vessel between bow and stern.

COMPASS ......Instrument for determining course steered and bearings.

COURSE ....Point of compass toward which vessel is steering.

DAVIT ....One of a pair of small cranes on a ship's side for hoisting and lowering lifeboats.

Gravity type.Davits which carry the lifeboat in two cradles on rollers which move on two parallel tracks to ship's side.

Radial type..Davits which carry the lifeboat on chocks and are crook-shaped; in launching, they are turned so that their tops describe an arc in placing the lifeboat over water from a position over the deck.

Quadrantal type ......Davits which carry the lifeboat on chocks, the davits themselves standing upright with the tops curved in toward each other so that the ends come directly above the hoisting hooks of the lifeboat.

DECK ....A platform or horizontal floor which extends from side to side of a vessel.

FALL ....A tackle of two blocks and length of rope for hoisting a boat to its davits.

FEND ....To push off when making a landing.

FLOAT ....See LIFE FLOAT.

FLOATING .....Act of being supported or buoyed by the water.

GUNWALE .....Upper edge of the side of an open boat.

HULL ....Frame or body of a vessel or boat, not including masting, rigging, etc.

KEEL ....Backbone of a ship or boat running from stem to sternpost at bottom of ship.

LANYARD .....Rope made fast to an article for securing it.

LASH ....To tie or secure.

LASHING ......A fastening made by passing a rope, cord, or the like around two or more objects, to hold one to the other.

LATITUDE .....Distance on the earth's surface north or south of the Equator.

LAUNCH ....To place in the water.

LEEWARD .....The direction toward which the wind blows; sheltered side.

LEEWAY ...... Drift of a vessel or boat to leeward caused by the wind or tide.

LIFEBOAT ..... A ship's boat particularly adapted and secured for use during an emergency.

Rubber type .. See CARBON DIOXIDE RUBBER BOAT.

Self-bailing type ...... Lifeboat which has a watertight compartment running its length where leaks are taken care of by handpumps.

Self-righting type ...... Lifeboat with high buoyancy (heavy keel, high air tanks, etc.).

LIFE BUOY .... See LIFE RING.

LIFE FLOAT .. Emergency buoyancy equipment of at least 15-person capacity that is launched overboard or floats free of a sinking ship.

LIFE-LINE .... Rope running length of sides of lifeboat for those in water to grasp.

LIFE NET ..... Square rope net used for slinging cargo and employed as a broad ladder for disembarking into landing craft or for abandoning ship.

LIFE RAFT .... Life float constructed with a metallic tube covered with cork and canvas or made of balsa wood or other suitable material. Life rafts are in four sizes: 67-, 41-, 28-, and 20-person capacity.

LIFE RING ....A buoyant device, usually in the shape of a ring, for keeping a person afloat.

LONGITUDE ...Distance east or west on earth's surface.

LUBBER LINE..Vertical line marked on inner surface of bowl of compass indicating the compass direction of a ship's head.

MANHOLE .....An opening into a tank or compartment.

MARLINE ......A small rope of two strands loosely twisted together; used for lashing.

MARLINSPIKE .A sharp-pointed iron pin used in splicing ropes.

PAINTER ......Line in the bow of a boat for towing or making fast.

PORT ....Left side of ship facing forward; opening in ship's side; a harbor.

RIG ....To fit out; general description of vessel's superstructure.

RIGGING ......Ropes or lines securing masts, booms, or sails.

ROWLOCKS ....Forked pieces of metal, fitted into sockets on rail of boat, in which the oar rests.

RUBBER BOAT.See CARBON DIOXIDE RUBBER BOAT.

RUDDER ....Flat vertical structure attached to stern of vessel for steering.

SAILMAKER'S PALM ....A leather band with a lead shield, fitting over the hand with shield in the palm; used as a thimble to push needle through canvas or rope.

SECURE ....To make fast; safe.

SHIP ....Vessel of large size carrying boats.

SHOAL ....A shallow place in any body of water; a sandbank or bar.

STANCHION ...Wooden or metal uprights used for support.

STARBOARD ...Right side of a ship looking forward.

STEP ....Frame on the keelson into which heel of mast is fitted.

STERN ....After part of a vessel.

STRETCHER ...Foot brace for oarsmen set athwartship.

STRINGER .....Horizontal plank or plate secured to a vessel's frame and supporting beam ends.

TAIL-BLOCK ...Block with a tail of rope instead of a hook.

THOLE-PINS ...Pins in gunwale of boat used for rowlocks.

THWART ......Crosspiece used as seat in lifeboats.

TILLER ....Short piece of iron or wood fitting into the head of a rudder, by which the rudder is turned.

TRACK ....Path of a vessel.

WATER-BORNE.Clear of the bottom.

WEATHER .....To windward; exposed to wind and rain.

WINDWARD ...In the direction from which the wind is blowing.

YOKE . . . . Athwartship piece fitting over rudderhead by which rudder is moved by means of ropes when tiller is not shipped.

☆ U. S. GOVERNMENT PRINTING OFFICE: 1950 — 883473

Made in United States
Troutdale, OR
11/03/2026

56844659R00094